Guide to Fossil Man

A Handbook of Human Palaeontology

BY MICHAEL H. DAY

Professor of Anatomy
St. Thomas's Hospital Medical School
University of London

With a Foreword by
PROFESSOR J. S. WEINER

THIRD EDITION
Completely revised and enlarged

CASSELL
LONDON

To Micky

for her support and companionship through swamps and deserts, in museums and collections, but most of all at home at the end of the day

CASSELL & COMPANY LIMITED
35 Red Lion Square, London WCIR 4SG
and at Sydney, Auckland, Toronto, Johannesburg,
an affiliate of
Macmillan Publishing Co., Inc.,
New York

First edition November 1965
Second edition (with corrections) April 1967
Third revised edition 1977

ISBN O 304 29835 2 (cloth)
ISBN O 304 29949 9 (paperback)

Printed in Great Britain by
The Camelot Press Ltd, Southampton

Foreword to Third Edition

What were our ancestors really like? How far back can we trace them? Can we discern in the fossil record a continuous sequence of transformation? Put simply, questions such as these constitute tests of the theory of evolution as applied to mankind. To answer such questions demands an understanding both of evolutionary theory and a detailed knowledge of the evidence—the fossil material. The fossil evidence after a century of remarkable discovery—and discovery at an accelerating pace—is now formidable in quantity, of challenging complexity, and spread over a forbidding number of monographs and scientific papers. To be provided with an accessible, comprehensive, up to date, reliable yet compact Guide to this wealth of information is indeed a matter for rejoicing. As I wrote in the preface to the first edition, this is precisely what Professor Michael Day has accomplished in this Guide and for which all students of the subject have been and will continue to be grateful.

Most students, even professionals, will never be in a position to examine complete collections of the casts available (let alone the original material) or to scrutinize at first hand the faunistic and geochronological evidence. They can, nevertheless, turn with confidence to this Guide for a soundly based and detailed appreciation of the key fossil material. The Guide provides a judicious coverage of all the important finds, each described in sufficient detail and illustrated by excellent photographs and accompanied by information on associated tools and fauna, and on the likely dating. In addition, Professor Day gives an appraisal of each find to enable the reader to appreciate its taxonomic and phylogenetic significance within the fossil record as a whole. His analysis errs on the side of caution—a conspicuous virtue in a subject shot through with dogmatism and controversy. This is a Guide both to the material and its interpretation.

In the preface to the first edition I pointed out that 'many of the fossil "men" we talk about so far from being known from whole skeletons are very sadly incomplete'. Parts of the evolutionary record are, unfortunately, still badly deficient but there has been a remarkable overall improvement. As Professor Day points out, over 500 finds of hominid remains have been recorded in the past eighteen years, and the majority of these since he produced the first edition. It is not only the older sites of East and South Africa which have continued to be productive but many new sites in Kenya, Ethiopia, Greece, France and Israel have yielded important and exciting finds.

Professor Day has kept fully abreast of all this new material, much of which he knows at first hand. For his dedication in making available, once again in the form of an accessible Guide, his unrivalled knowledge of the fossil evidence of man's evolution, we are grateful.

J. S. Weiner
Past President
Royal Anthropological Institute

Contents

List of Illustrations

Within recent years the subject of human palaeontology, the study of fossil man, has experienced an avalanche of new material from a variety of sites, particularly in Africa but also from other parts of the world. This wealth of new primary data, while welcome, has brought unprecedented problems by the sheer volume of material for evaluation. Over 500 finds of hominid remains have been recorded in the past eighteen years and the majority of these within the last seven years. Sites such as Koobi Fora (formerly East Rudolf, Kenya), Hadar (Ethiopia), Laetolil (Tanzania), Petralona (Greece), Arago (France), Amud and Jebel Kafzeh (Israel) and Jebel Ighoud (Morocco) have all produced valuable and significant new finds. Some longer-known sites such as Olduvai Gorge, Swartkrans, Sterkfontein and Sangiran, are still productive. It is not surprising that in this situation there is more than a little confusion in the interpretation of the finds since the rate of discovery has outpaced the work of analysis and the gap has widened between publication, monographic treatment and inclusion in general texts. The task of eager students following courses in human evolutionary studies has been made even more difficult than before in selecting the sites and the specimens about which they should know. From this point of view the aim of this book remains what it was twelve years ago 'to bring together information from many disciplines and relate it for each of the 40 or so key sites which have yielded the bulk of significant hominid fossils'. Of necessity some sites have been dropped and others have come to take their place, not only because they are new but also because they are better dated, or have better material, or because they extend the range of our knowledge of hominid evolution.

In order to accommodate as great a range of sites as possible it has been decided to drop the general sections on the environment, culture and dating since much of this information is now incorporated in the site description while the remainder is better treated in other specialist books published in recent years.

A number of new field and laboratory techniques have been added to the armamentarium of the palaeoanthropologist, mostly by physicists; these include refinements of the potassium-argon dating method, the development of the fission-track dating method, amino-acid racemization dating and the advent of geomagnetic polarity determination. The latter technique is providing valuable checks on stratigraphy by permitting comparisons with the world geomagnetic polarity column. Some of these methods are still in their infancy but promise well for the future.

Anatomical analysis has continued to be the backbone of hominid fossil interpretation but the use of multivariate statistical analysis of osteometric data has not proved to be as useful a method as had been hoped since the biological interpretations put upon the results obtained have often been confusing. It remains clear, however, that statistical significance and biological significance are not necessarily to be equated and that without anatomical analysis statistical deductions are an inadequate basis upon which to make taxonomic or locomotor judgements. Indeed, the future of locomotor studies would seem to be with biomechanics rather than statistics.

The study of fossil man still attracts widespread public attention in the news media, in documentary films, and in popular books. This interest is valuable and desirable but it imposes upon all workers in the field a duty, and this is to be as accurate as possible. While I am indebted to many for assistance, the errors are all my own.

February 1977

Introduction to the First Edition

The study of human evolution presents many problems of peculiar difficulty, perhaps the greatest being the large and increasing number of specimens forming the evidence upon which the subject is based. In this book an attempt has been made to select, from the wide range of known material, the hominid fossils that best illustrate a significant stage or aspect of human palaeontology, the study of fossil man.

Anatomical examination of fossil bones and teeth can be revealing, tantalizing, exasperating. Always the temptation is to speculate beyond the facts, to argue from the particular to the general, to forget that one specimen represents a population and is therefore subject to the laws of biological variation. However, if these pitfalls are avoided, there remains an ever increasing body of knowledge of the anatomy of fossil man. Even so, anatomical knowledge is not enough upon which to base an assessment of the life of early man; other factors must be considered, the climate, the time relationships of different groups and the culture of the populations represented by the scant remains at our disposal.

It is the purpose of this book to bring together information from many disciplines and relate it for each of the 40 or so key sites which have yielded the bulk of significant hominid fossils.

November, 1965

Acknowledgments

It is a great pleasure to record my most grateful thanks to all of those who have helped me during the preparation of the third edition of this book. I am particularly grateful to Mrs Berna Lindfield, my Research Assistant, who has cheerfully borne the brunt of the library work and the correspondence.

For the opportunity to examine original material, the provision of photographs and for valuable discussions I am indebted to many, but I must mention the late Professor C. Arambourg, the late Professor W. W. Bishop, Dr C. K. Brain, Mr R. J. Clarke, Dr D. A. Hooijer, Dr F. C. Howell, Dr W. W. Howells, Dr G. L. Isaac, Dr D. C. Johanson, Dr G. H. R. von Koenigswald, Dr R. Kraatz, the late Dr L. S. B. Leakey, Dr M. D. Leakey, Dr M. G. Leakey, Mr R. E. F. Leakey, Professor H. de Lumley, Dr M. de Lumley, Miss T. I. Molleson, Professor J. R. Napier, Dr J. J. Oberholzer, Dr J. T. Robinson, Dr R. Singer, Dr T. D. Stewart, Dr H. Suzuki, Dr C. B. Stringer, Dr A. T. Sutcliffe, Dr A. Thoma, Professor P. V. Tobias, Dr B. Vandermeersch, Dr A. C. Walker, Professor J. S. Weiner, Dr J. K. Woo, Dr B. A. Wood, the Director of the Musée de l'Homme, Paris, the Trustees of the National Museums of Kenya, and the Trustees of the British Museum (Natural History).

I must also thank most kindly the Librarians of St Thomas's Hospital Medical School and my former secretary Mrs Ann Capps for their unfailing help. The text-figures and maps reflect the skill of Mrs Audrey Besterman.

PART I

The principal evidence upon which direct knowledge of the evolution of man rests has been provided by anatomical examination of fossil bones and teeth. Although skeletal and dental structure may appear limited sources of information, behavioural characteristics—such as locomotor capability and dietary habit—may be deducible from their study.

In the past, emphasis was placed upon the phyletic value of single features whose functional significance was often obscure, for example, the variations in the arrangement of skull sutures. However, wider knowledge of the range of variation of single features in all primates has allowed these characteristics to be regarded in proper perspective.

The study of individual bones is of greater value, and an analysis of joint geometry and muscular mechanics can often strongly infer joint function. When there is available a group of bones that form a functional complex (hand, foot, pelvis, vertebral column, jaws and teeth) its examination will give the beginnings of an insight into the life of the form of which it once formed part. Also, with sufficient material, knowledge of several functional complexes may be built up to provide almost the entire picture, morphological and functional; such a process, however, must stop short of modelling soft tissue features and hair distribution, for which no direct evidence exists.

Bones

It is apparent that the study of fossil man requires a working knowledge of the skeleton of modern man; the terminology of human palaeontology closely parallels that of human osteology, but it is not within the scope of this book to describe the human skeleton, a topic dealt with in all standard anatomical texts. However, perhaps certain aspects of osteology acquire importance when set against a comparative background, and thus deserve emphasis.

Bone is living tissue, if it is cut, it bleeds; if broken, it heals. As with other living tissue, it reacts to stimuli; in particular it responds to mechanical stresses, a principle embodied in Wolff's Law:

> 'The external form and internal architecture of a bone are related to the forces which act upon it.'

It follows that the shape of a bone, although genetically determined, 3 is capable of modification, and at death the external form and internal architecture will reflect the work that it was called upon to perform during the latter part of life. It is upon this premise that much of the interpretation of skeletal morphology depends.

NORMAL SKELETAL VARIATION

It is a commonplace that no two people are alike; neither are their skeletons nor the individual bones and teeth of which they are composed. Every morphological feature and every measurement forms but a part of the range of normal biological variation to be found within a population. Sampling techniques can be employed to define the approximate limits of this range and the distribution of characteristics within it. While statistical methods can assist in the evaluation of observations and in the comparison of data, unfortunately hominid fossil bones and teeth are frequently too few in number to allow proper statistical comparisons between them and samples drawn from populations of related forms. Often the most that can be achieved is to determine whether or not a given character (which may itself be at any point within its own range of normal variation) lies within the range of normal variation of that character in an allied form.

In order to facilitate comparison, indices can be devised which allow two variables to be expressed as a single figure; length in terms of breadth (crania, tooth crowns), shaft thickness in terms of length (long bones). Comparison of indices allows specimens of differing absolute sizes to be compared in respect of a relative feature. Increasingly the more sophisticated statistical techniques of multivariate analysis are being used in attempts to analyse the relationships between groups of bone measurements. The advent of electronic computers and package programs has facilitated this development.

In essence most of the methods involve an attempt to reduce the comparative data to manageable proportions and often they permit a visualization of the relationships of the tested groups by means of graphic methods in two or more dimensions. Part of the difficulty of using these methods stems from a realization that the mathematical model is multi-dimensional so that a simple visualization is not really possible and, if attempted, can be misleading. The statistics that are achieved (e.g. Hotelling's T^2, Mahalanobis' D^2 and others) may be regarded as measures of morphometric distance, and even be assigned levels of significance, yet their true biological significance often remains obscure and the subject of controversy. This is particularly true when such measures are used as evidence in the taxonomic evaluations of scanty fossil remains, themselves often reconstructed prior to measurement. Data drawn from limb bones may be expected to produce evidence that will relate to locomotor function but the use of such evidence in a taxonomic evaluation may in turn depend upon the validity of locomotor evidence in taxonomy.

At best statistical evidence can only complement anatomical evidence and this applies only *when the measurements are taken directly from the original fossils* and due allowance is made for speculative reconstructions and distortion. In the evaluation of fossil hominid remains nothing replaces an intimate familiarity with the anatomy of the specimens gained by study of the originals. Casts and photographs have a place but they should never be used as primary sources for data in serious work. It follows that published work that does not specify that it is based upon examination of originals should be regarded with caution.

SEXUAL VARIATION

Sexual dimorphism is often regarded as a common feature of the skeletons of the hominids including modern man, and on this account it is frequently taken for granted that a large robust bone is male whereas a small gracile bone is female. A moment's reflection should suffice to establish that this is not an invariable rule in modern man. What is probably true is that the means of the ranges of normal skeletal size variation of male and female hominids are separated, the male being the higher; however, this gives no firm basis for asserting the sex of any individual unknown hominid bone, fossil or recent, on the grounds of size.

The only sexual characters upon which some reliance may be placed are the morphological features of the pelvis; it is well known in modern man that these will assist in differentiating male from female pelves (see Table 1).

Table 1   The Principal Differences between the Modern Human Male and Female Pelvis

MALE PELVIS	FEMALE PELVIS
1. Sacrum straight and upright	Sacrum hollowed and forward tilted
2. Inferior pubic ramus strongly everted	Inferior pubic ramus weakly everted
3. Sciatic notch 90° or less	Sciatic notch 90° or more
4. Inferior pubic angle 90° or less	Inferior pubic angle 90° or more
5. Length of superior pubic ramus equals acetabular diameter	Length of superior pubic ramus is greater than acetabular diameter
6. The ilia are upstanding	The ilia are flattened
7. The ischial spines are prominent and project inwards	The ischial spines are not prominent and do not project inwards

None of the characters listed above is diagnostic on its own, but taken in conjunction with each other they provide some evidence upon which to base an opinion. The relationships that exist between bone size and bone shape are now known to be of increasing importance and the study of these relationships, bone allometry, is receiving considerable attention in the current literature. More information about allometric relationships in turn is throwing light upon sexual dimorphism.

AGE VARIATION

Growing Phase. Age changes in bones are well recognized during the period of skeletal development. Long bones ossify from centres which appear early in intra-uterine development near the mid-point of the cartilage pre-cursor of the shaft. The ossification process spreads from this diaphysial centre towards the ends of the bone; later new epiphysial centres appear at the ends. The cartilage plate separating the diaphysis from the epiphysis constitutes the growing zone. Additional bone is laid down beneath the periosteum which surrounds the shaft. Remodelling, by resorption and addition, modifies the structure until development is nearly complete, then the epiphysis fuses on to the shaft and the bone is adult.

The state of skeletal maturity allows some estimate to be made of age, although variability is common between individuals and between species. It is probably unwise to make estimates of the chronological age at death of an immature fossil bone on the basis of correlation with similar bones of modern man, other than in broad categories such as foetal, infant, juvenile and adolescent.

Mature Phase. During adult life age changes are comparatively few. Tooth attrition is continuous, but its rate is influenced by diet and the state of dental health. Typically, adult fossil hominid incisors are worn down so that in occlusion the bite is 'edge to edge'.

Skeletal changes during adult life include the gradual obliteration of the cranial sutures by fusion (synostosis) of adjacent bones. The sequence of closure has long been regarded as giving a guide to chronological age; however, it is in fact so unreliable in modern man that it is unwise to attempt to determine the age of an unknown fossil hominid by means of correlation, even if there were grounds for suggesting that such a correlation exists.

Senile Phase. The changes of senility are degenerative; they include osteoarthritis, osteoporosis and changes in the vertebral column which produce bent posture and loss of stature. Loss of teeth is sometimes

Table 2   *Modern Human Post-Cranial Skeletal Maturation (Long bones)*

| Bone | Appearance of Centres | | Fusion of Diaphysis and Epiphysis |
	Diaphysis	Epiphysis	
Clavicle	5–6/52 intrauterine life (i.u.)	c. 17–21	c. 20 +
Humerus	8/52 i.u.	Upper 6/12–5	18–20
		Lower 2–12	14–16 (Med. epicond. c. 20)
Radius	8/52 i.u.	Upper 4–5	14–17
		Lower 1	17–19
Ulna	8/52 i.u.	Upper 9–11	14–16
		Lower 5–6	17–18
Femur	7/52 i.u.	Upper 6/12–14	14–17
		Lower c. birth	16–18
Tibia	7/52 i.u.	Upper c. birth	16–18
		Lower 10/12	15–17
Fibula	8/52 i.u.	Upper 3–4	17–19
		Lower 1	15–17
Pelvis			
Ilium	8/52 i.u.	secondary centres c. puberty	Ischium–Pubis 7–8 Complete 15–25
Ischium	4/12 i.u.	secondary centres c. puberty	
Pubis	4–5/12 i.u.	secondary centres c. puberty	

After Warwick and Williams (1973)

associated with senility; complete tooth loss, with resorption of alveolar bone, considerably changes the stresses acting upon the mandible, and in turn leads to bone remodelling at the gonion and the temporo-mandibular joint.

However, since all of these 'senile' changes are pathological and may occur in individuals who are not old in a chronological sense, 'senile' skeletal or dental changes must be regarded cautiously as evidence of ageing in hominid fossil material.

Teeth

Tooth morphology has played an important part in vertebrate palaeontology, if only because teeth preserve well and dental characters are easily identified. The human dentition is described fully in textbooks of dental anatomy, but it may be of value to draw attention to some features of anthropological interest.

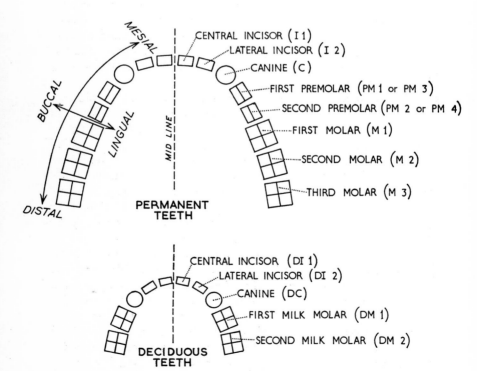

Fig. 1   The nomenclature of hominid upper and lower teeth

TERMINOLOGY

The hominid dental formula is the same as that of catarrhine monkeys and great apes.

Deciduous Teeth   Incisors (DI) $\frac{2}{2}$ Canines (DC) $\frac{1}{1}$ Molars (DM) $\frac{2}{2}$

Permanent Teeth   Incisors (I) $\frac{2}{2}$ Canines (C) $\frac{1}{1}$ Premolars (PM) $\frac{2}{2}$ Molars (M) $\frac{3}{3}$

That part of a tooth which projects above the gum is known as the crown; its surfaces are named mesial, distal, buccal, lingual and occlusal (Fig. 1).

In the case of premolars and molars, the biting or occlusal surface is modified by projections or cusps which are separated by fissures.

Premolar teeth are often referred to as 'third' or 'fourth' premolars, an indication that the Old World higher primate dental formula was derived from the generalized mammalian formula by loss of teeth and in particular by the loss of the first two premolars. Many publications refer to individual teeth by a shorthand method which indicates whether teeth are from the upper or lower dentition (e.g. PM_3 or $PM_{\overline{3}}$ meaning lower third premolar or M^1 or $M^{\underline{1}}$ meaning upper first molar).

The general nomenclature of mammalian molar crown morphology was developed by Cope and Osborn in the late nineteenth century, and enables any cusp or ridge to be identified. The primary distinction between upper and lower molar teeth is made by adding the suffix 'id' to the name of any part of a lower molar. This allows the same terms to apply to both upper and lower molar features yet remain clearly distinguishable. The basic structure is a three-cusped triangle or *Trigon* (*Trigonid*), whose apex is lingual in upper molars and buccal in lower molars. The apical cusp is the *Protocone* (*Protoconid*), the mesial cusp the *Paracone* (*Paraconid*) and the distal cusp the *Metacone* (*Metaconid*).

Distally the trigon (trigonid) is often extended by a basin-like *Talon* (*Talonid*) which carries a major cusp, the *Hypocone* (*Hypoconid*). The talonid frequently has a lingual *Entoconid* in addition to a hypoconid. Minor cusps which occur on ridges linking major cusps are distinguished by the suffix 'conule' (conulid), and are attributed to the nearest major cusp, for example *Protoconule* or *Hypoconulid*. A basal collar of enamel which may develop around a tooth is termed a cingulum (pl. cingula); minor cusps which arise from cingula are given the suffix 'style' (stylid) and are attributed to the related major cusp, for

9

example *Parastyle* or *Parastylid.*

Although this terminology may appear complex it has the virtue of permitting direct reference to the features of tooth crown morphology in a wide range of mammals.

HOMINID DENTAL CHARACTERS

The principal features of hominid dentition which distinguish it from that of the great apes are the regularly curved shape of the dental arcade, reduction in the size of incisors and canines, a non-sectorial first premolar, and premolar and molar teeth with a flat occlusal wear

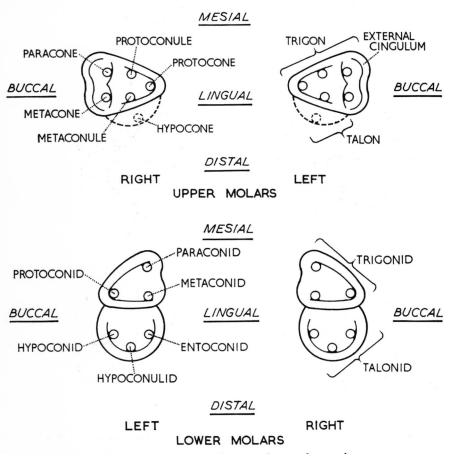

Fig. 2   The general nomenclature of mammalian molar tooth crown morphology (after Simpson, 1937)

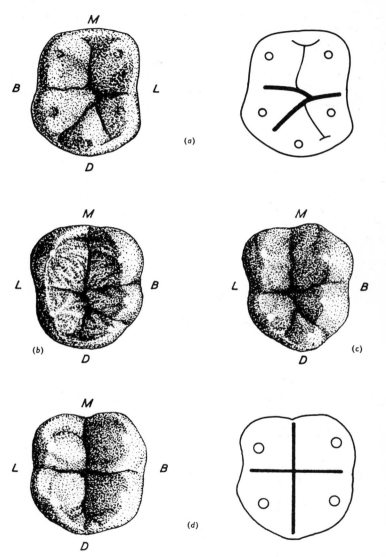

Fig. 3   The cusp and fissure pattern of pongid and hominid lower second
permanent molar teeth
(a) *Dryopithecus* (Left, Y5)   (b) Modern chimpanzee (Right, Y5)
(c) Modern man (Right, Y5)   (d) Modern man (Right, +4)

pattern. However, the hominoid lower molar cusp pattern is basically the same as that of a Miocene ape known as *Dryopithecus*; this arrangement consists of five cusps separated by grooves in the form of a Y (Y5), and is typical of modern apes. This persisting cusp and fissure pattern is commonly found in the lower molar teeth of fossil man, but in modern man there is frequently reduction or absence of the fifth cusp (hypoconulid) and the formation of a + fissure pattern (+5 or +4), particularly in the second and third molars.

TOOTH ERUPTION

The times of eruption of modern human teeth have long been known to give an indication of age. Unfortunately the ranges of normal variation are apt to be very wide so that estimations of age based on correlation within this species tend to be unreliable. It is even more hazardous to attempt to correlate the ages of tooth eruption of fossil hominids with those of modern man, other than in broad general terms.

The following table gives the approximate ages of tooth eruption for modern man.

Table 3   Approximate Ages of Tooth Eruption in Modern Man

Upper Deciduous Teeth — Months

Tooth	4	6	8	10	12	14	16	18	20	22	24	26	28	30	32	34	36
DI_1			■	■	■	■											
DI_2				■	■	■	■										
DC							■	■	■	■							
DM_1					■	■	■	■	■								
DM_2										■	■	■	■	■	■	■	■

Lower Deciduous Teeth — Months

Tooth	4	6	8	10	12	14	16	18	20	22	24	26	28	30	32	34	36
DI_1		■	■	■													
DI_2				■	■	■	■	■									
DC							■	■	■	■	■						
DM_1					■	■	■	■									
DM_2										■	■	■	■	■	■	■	■

Upper Permanent Teeth — Years

Tooth	4	5	6	7	8	9	10	11	12	13	14	15	16	17	18	19	20	21	22	23
I_1			■	■	■															
I_2			■	■	■	■														
C				■	■	■	■													
PM_1				■	■	■	■	■	■											
PM_2				■	■	■	■	■	■	■										
M_1		■	■	■																
M_2						■	■	■	■	■										
M_3												■	■	■	■	■	■	■	■	■

Lower Permanent Teeth — Years

Tooth	4	5	6	7	8	9	10	11	12	13	14	15	16	17	18	19	20	21	22	23
I_1			■	■	■															
I_2			■	■	■	■														
C					■	■	■	■	■											
PM_1				■	■	■	■	■	■											
PM_2				■	■	■	■	■	■	■										
M_1		■	■	■																
M_2						■	■	■	■	■										
M_3												■	■	■	■	■	■	■	■	■

References and Further Reading

SIMPSON, G. G. 1937 'The beginning of the age of mammals'. *Biol. Rev.* *12*, 1-47.

BREATHNACH, A. S. Ed. 1958 Frazer's *Anatomy of the Human Skeleton*. 5th Ed. London: Churchill.

MARTIN, R. and SALLER, K. 1959 *Lehrbuch der Anthropologie*. Stuttgart: Gustav Fischer Verlag.

SCOTT, J. H. and SYMONS, N. B. B. 1961 *Introduction to Dental Anatomy*. Edinburgh and London: E. and S. Livingstone.

KROGMAN, W. M. 1962 *The Human Skeleton in Forensic Medicine*. Springfield, Illinois: Charles C. Thomas.

ROMANES, G. J. Ed. 1964 *Cunningham's Textbook of Anatomy*. 10th Ed. Oxford: Oxford University Press.

OLIVIER, G. 1965 *Anatomie Anthropologique*. Paris: Vigot Frères.

ROBINSON, J. T. and ALLIN, E. F. 1966 'On the Y of the Dryopithecus Pattern of Mandibular Molar Teeth'. *Am. J. Phys. Anthrop.* *25*, 323-324.

WARWICK, R. and WILLIAMS, P. L. 1973 *Gray's Anatomy*. 35th Ed. London: Longmans.

PART II

The Fossil Hominids

The Name of the Finds

COUNTRY OF ORIGIN   REGION

Synonyms and other names The zoological names given to the specimens by other authors are quoted with the appropriate reference; in addition colloquial names in common use will be mentioned. Names given under this heading are not necessarily strict taxonomic synonyms.

Site The location of the find. Distances are given in miles or kilometres according to the original publications.

Found by The name of the field worker responsible for the find and the date.

Geology A summary of the geology and stratigraphy of the site.

Associated finds The artefacts and fossil fauna found with the remains.

Dating An assessment of the dating of the specimens.

Morphology A description of the specimens.

Dimensions Selected dimensions are given, from measurement of the original bones, from published works and, on occasion, from measurement of casts. All measurements are in millimetres unless otherwise indicated.

Affinities A summary of opinions on the nomenclature, classification and relationships of the finds.

Original The address of the institution where the remains are usually kept.

Casts Where possible the addresses are given of concerns or institutions where at one time or another casts might be obtained. No assurance can be given that casts are currently available from any of the sources mentioned.

References The principal references to the literature concerning the find.

16

Europe and the Near East

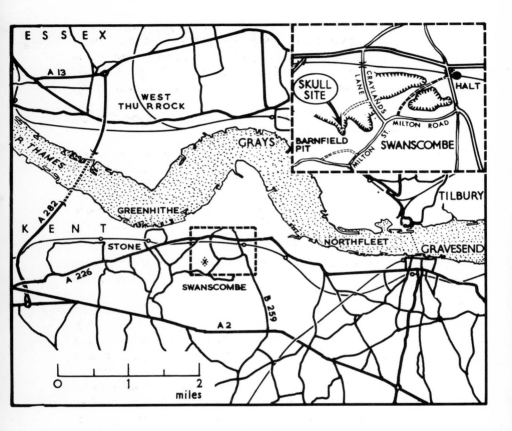

Fig. 4　The Swanscombe site

The Swanscombe Skull Fragments

Synonyms and other names — *Homo cf. sapiens* (Le Gros Clark *et al.*, 1938); Praesapiens man (Vallois, 1954); *Homo sapiens protosapiens* (Montandon, 1943); *Homo marstoni* (Paterson, 1940); *Homo swanscombensis* (Kennard, 1942); *Homo sapiens steinheimensis* (Campbell, 1964); Swanscombe man (Oakley, Campbell and Molleson, 1971).

Site — Barnfield pit, half a mile south-west of All Saints Church, Swanscombe, Kent.

Found by — A. T. Marston, 29th June, 1935, and 15th March, 1936; Mr and Mrs B. O. Wymer, A. Gibson and J. Wymer, 30th July, 1955.

Geology — The bones were found in river gravel 24 feet below the surface of the 100-foot terrace of the Thames in an oblique bed at the base of the Upper Middle Gravel.

Associated finds — The implements found with the remains include numerous hand-axes and flake tools, an industry equivalent to the Middle Acheulean culture (Breuil, Acheulean III).

The fossil mammalian fauna recovered from the Swanscombe interglacial deposits includes 26 species (Sutcliffe, 1964). Amongst those recognized from the Upper Middle Gravel are wolf (*Canis cf. lupus*), lion (*Panthera spelaea*), straight tusked elephant (*Palaeoloxodon antiquus*), Merck's rhinoceros (*Didermocerus kirchbergensis = mercki*), horse (*Equus cf. caballus*), fallow deer (*Dama clactoniana*), giant deer (*Megaceros giganteus*), red deer (*Cervus elaphus*), giant ox (*Bos primigenius*) and hare (*Lepus sp.*). During 1968, 1969 and 1970 new excavations have been conducted at Swanscombe (Waechter *et al.* 1969–1972) into the Lower Loam and Lower Gravels. Worked flint and mammalian bones have been recovered.

The most significant finds were retrieved at the junction between the Lower Loam and Lower Gravels on what has been termed a 'floor'. On this floor was a skull of cave bear (*Ursus spelaeus*) and several bovid and cervid fragments.

Dating — On geological and faunal grounds, there seems little doubt that the deposit dates from the close of the Second Interglacial (Mindel–Riss or Penultimate Interglacial), a date which

19

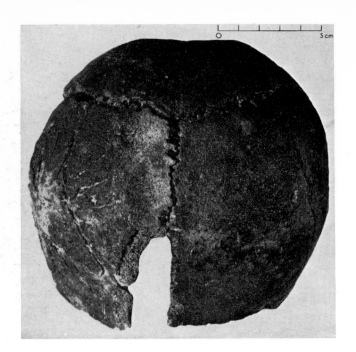

Fig. 5

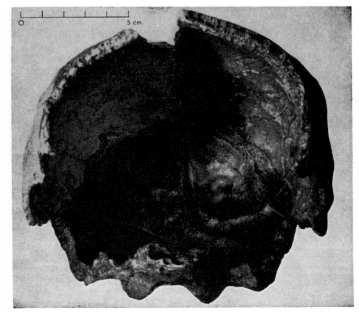

Fig. 6

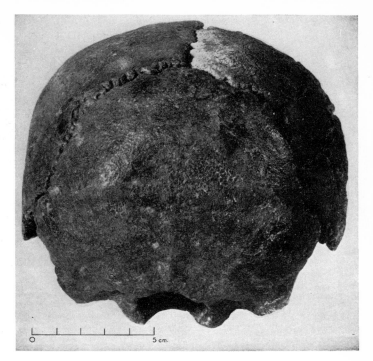

Fig. 7

Fig. 5   The Swanscombe skull
         Vertical view
         Courtesy of the Trustees of the British Museum (Nat. Hist.)

Fig. 6   The Swanscombe skull
         Internal view
         Courtesy of the Trustees of the British Museum (Nat. Hist.)

Fig. 7   The Swanscombe skull
         Occipital view
         Courtesy of the Trustees of the British Museum (Nat. Hist.)

is confirmed by the type of tool culture associated with the remains.

No pollen was found in the deposits, but the fluorine content of the skull fragments and the associated mammalian bones is similar, confirming the contemporaneity of the specimens (Oakley, 1949). More recently the Second Interglacial date of the Swanscombe fragments has been reaffirmed by Kurtén (1962), Oakley (1964), and Szabo and Collins (1975).

Morphology   The bones comprise an occipital (1935), a left parietal (1936) and a right parietal (1955). All three are almost complete and undistorted, the sutures are open and the bones clearly fit together; it is believed that they belong to a young adult. In general the bones are of modern form, but the parietals are exceptionally thick. The occipital shows no sign of a chignon and the transverse ridge—scarcely a torus—is more prominent at its ends than in its central portion. The occipital condyles and the orientation of the foramen magnum do not differ significantly from those features in *Homo sapiens*. The sphenoidal sinus must have been large as it extends into the basi-occiput. The parietals are quadrangular and marked by the temporal lines which are well separated from each other and pass below the parietal eminence. The parietal foramina are absent and the biasterionic breadth is large. Neither parietal has a Sylvian crest on its internal surface, but the meningeal vascular pattern is complex and of unusual form by comparison with modern human bones. The grooves for the dural venous sinuses are asymmetrical.

Dimensions   *Swanscombe Committee* (1938). *Morant*
Max. Length (181·5)   Max. Breadth (142)
Biasterionic Breadth 123·5?   Cranial Index 78?
Cranial Capacity (1,325 cc)   (Mesocephalic?)

Weiner and Campbell (1964)
Biasterionic Breadth 123·0   Max. Breadth (145)

? Close approximation
( ) Estimated

Affinities   The first two bones were studied by the Swanscombe Committee (Le Gros Clark *et al.*, 1938). In their report Morant showed that the metrical characters of the occipital and left   22 parietal bones could not be distinguished from those of *Homo sapiens* with the exception of the biasterionic breadth; because of this difference and the thickness of the vault Morant suggested that the frontal bone might be like that of Steinheim man. This view has been re-expressed by Breitinger (1952 and 1955). Vallois (1954) saw no reason to suggest that Swanscombe man had brow ridges and regards Swanscombe and Fontéchevade as 'Praesapiens' forms evolving in parallel with the Steinheim–Neanderthal line. Le Gros Clark (1955) has again drawn attention to the similarities between the Swanscombe and Steinheim forms, referring to both as representatives of 'Pre-Mousterian and Early Mousterian *Homo sapiens*'. Other authors are less convinced of the sapient features of Swanscombe man and emphasize the Neanderthal and primitive features of these remains (Breitinger, 1952 and 1955; Stewart, 1960). Howell (1960) affirmed that the known fragments from Swanscombe and Steinheim indicate that both forms represent the same hominid variety and that they 'deviate from the anatomically modern morphological pattern and, instead, are allied with early Neanderthal peoples'.

The Swanscombe site and fossil material have been thoroughly reinvestigated recently (Brothwell *et al.*, 1964). Following a morphological, metrical and statistical reappraisal of the skull bones, Weiner and Campbell (1964) emphasized the degree of interrelatedness between the forms of *Homo*, i.e. Solo, Rhodesian, Neanderthal and modern man. They suggest that it is not possible to maintain, in a taxonomic sense, strictly specific status for each of these forms, regarding them as a 'spectrum' of varieties within one species. Swanscombe man has been regarded as belonging to the Neanderthaloid 'intermediate' group which contains the Steinheim, Ehringsdorf, Skhūl V and Krapina specimens.

In an appraisal of the Vértesszöllös occipital (*q.v.*) Wolpoff (1971) has suggested that the Swanscombe skull probably represents a later female example of the same lineage. A lineage that he attributes to *Homo erectus*.

Originals   British Museum (Natural History), Cromwell Road,
South Kensington, London, S.W.7.   23

The Swanscombe Skull Fragments

Casts   The University Museum, University of Pennsylvania,
Philadelphia 4, Pennsylvania, U.S.A.

References   SMITH, R. A. and DEWEY, H. 1913
Stratification at Swanscombe. *Archaeologia 64*, 177–204.
MARSTON, A. T. 1936
Preliminary note on a new fossil human skull from Swanscombe,
Kent. *Nature 138*, 200–201.
MARSTON, A. T. 1937
The Swanscombe skull. *J. R. Anthrop. Inst. 67*, 339–406.
CLARK, W. E. LE GROS, OAKLEY, K. P., MORANT, G. M., KING, W. B. R.,
HAWKES, C. F. C., *et. al.* 1938
Report of the Swanscombe Committee. *J. R. Anthrop. Inst. 68*, 17–98.
PATERSON, T. T. 1940
Geology and early man. *Nature 146*, 12–15, 49–52.
KENNARD, A. S. 1942
Faunas of the High Terrace at Swanscombe. *Proc. Geol. Ass. Lond.
53*, 105.
MONTANDON, G. 1943
L'homme préhistorique et les préhumains. Paris: Payot.
OAKLEY, K. P., and MONTAGU, M. F. A. 1949
A reconsideration of the Galley Hill skeleton. *Bull. Brit. Mus. (Nat.
Hist.) 1*, 27–46.
BREITINGER, E. 1952
Zur Morphologie und systematischer Stellung des Schädelfragmentes
von Swanscombe. *Homo 3*, 131–133.
OAKLEY, K. P. 1952
Swanscombe man. *Proc. Geol. Ass. Lond. 63*, 271–300.
VALLOIS, H. V. 1954
Neandertals and praesapiens. *J. R. Anthrop. Inst. 84*, 111–130.
BREITINGER, E. *1955*
Das Schädelfragmentes von Swanscombe und das 'Praesapiens-
problem'. *Mitt. anthrop. Ges. Wien 84/85*, 1–45.
CLARK, W. E. LE GROS 1955
The fossil evidence for human evolution. Chicago: Chicago University
Press.
WYMER, J. 1955
A further fragment of the Swanscombe skull. *Nature 176*, 426–427
OAKLEY, K. P. 1957
Stratigraphical age of the Swanscombe skull. *Am. J. Phys. Anthrop. 15*,
253–260.
STEWART, T. D. 1960
Indirect evidence of the primitiveness of the Swanscombe skull.
Am. J. Phys. Anthrop. 18, 363.
HOWELL, F. C. 1960
European and northwest African Middle Pleistocene hominids. *Curr.
Anthrop. 1*, 195–232

KURTÉN, B. 1962
The relative ages of the australopithecines of Transvaal and the pithecanthropines of Java. *In Evolution und Hominisation*, 74–80. Ed. G. Kurth. Stuttgart: Gustav Fischer Verlag.

BROTHWELL, D. R., CAMPBELL, B. G., CASTELL, C. P., GARDINER, E., OAKLEY, K. P., PATTERSON, C., SUTCLIFFE, A. J., SWINTON, W. E., WEINER, J. S., WYMER, J. 1964.
In *The Swanscombe Skull*. Ed. C. D. Ovey. Royal Anthropological Institute, *Occasional Paper No. 20*. London: Royal Anthropological Institute.

CONWAY, B. W. 1969
Preliminary geological investigation of Boyn Hill terrace deposits at Barnfield pit, Swanscombe, Kent, during 1968. *Proc. R. Anthrop. Inst. of Gt. Britain and Ireland for 1968*, 59–61.

WAECHTER, J. D'A. 1969
Swanscombe 1968. *Proc. R. Anthrop. Inst. of Gt. Britain and Ireland for 1968*, 53–61.

CONWAY, B. W. 1970
Geological investigation of Boyn Hill terrace deposits at Barnfield pit, Swanscombe, Kent, during 1969. *Proc. R. Anthrop. Inst. of Gt. Britain and Ireland for 1969*, 90–92.

NEWCOMER, M. H. 1970
The method of excavation at Barnfield pit, Swanscombe (1969). *Proc. R. Anthrop. Inst. of Gt. Britain and Ireland for 1969*, 87–89.

WAECHTER, J. D'A. 1970
Swanscombe 1969. *Proc. R. Anthrop. Inst. of Gt. Britain and Ireland for 1969*, 83–85.

CONWAY, B. W. 1971
Geological investigation of Boyn Hill terrace deposits at Barnfield pit, Swanscombe, Kent, during 1970. *Proc. R. Anthrop. Inst. of Gt. Britain and Ireland for 1970*, 60–63.

NEWCOMER, M. H. 1971
Conjoined flakes from the Lower Loam, Barnfield pit, Swanscombe (1970). *Proc. R. Anthrop. Inst. of Gt. Britain and Ireland for 1970*, 51–59.

OAKLEY, K. P., CAMPBELL, B. G. and MOLLESON, T. I. 1971
Catalogue of Fossil Hominids Part II: Europe. London: Trustees of the British Museum (Natural History).

WAECHTER, J. D'A. 1971
Swanscombe 1970. *Proc. R. Anthrop. Inst. of Gt. Britain and Ireland for 1970*, 43–49.

WOLPOFF, M. H. 1971
Is Vértesszöllös an occipital of *Homo erectus? Nature* 232, 867–868.

CONWAY, B. W. 1972
Geological investigation of the Boyn Hill terrace deposits at Barnfield pit, Swanscombe, Kent, during 1971. *Proc. R. Anthrop. Inst. of Gt. Britain and Ireland for 1971*, 80–85.

The Swanscombe Skull Fragments

WAECHTER, J. D'A. 1972
Swanscombe 1971. *Proc. R. Anthrop. Inst. of Gt. Britain and Ireland for 1971*, 73–78.
SZABO, B. J. and COLLINS, D. 1975
Ages of Fossil Bones from British Interglacial Sites. *Nature* 254, 680–682.

The La Chapelle-aux-Saints Skeleton

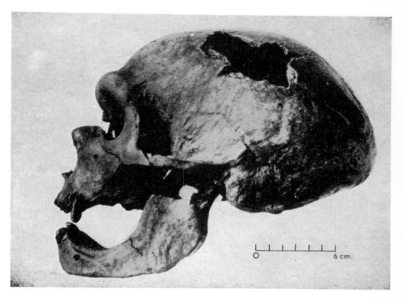

Fig. 8   The skull from La Chapelle-aux-Saints
Courtesy of the Director of the Musée de l'Homme

Synonyms | *Homo neanderthalensis* (Boule, 1911–1913); *Homo sapiens*
and other names | *neanderthalensis* (Campbell, 1964); Neanderthal man; Nean-
dertal man; La Chapelle-aux-Saints 1. (Oakley, Campbell
and Molleson, 1971).

Site | Near the village of La Chapelle-aux-Saints, 25 miles south-
east of Brive, Corrèze, France.

Found by | A. and J. Bouyssonie and L. Bardon, 3rd August, 1908.

Geology | The skeleton was found buried in the floor of a small cave
hollowed into the limestone of the Lower Lias, which rests
upon Triassic sandstone. The skeleton was covered with
calcareous clay containing stones which had fallen from the
roof. The stratigraphy was carefully recorded whilst the
remains were uncovered.

Associated finds | With the bones were numerous flint tools of an evolved
Mousterian culture including retouched blades, scrapers and

27

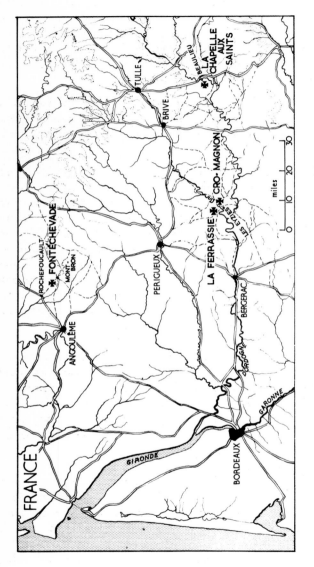

Fig. 9   Hominid fossil sites in southern France

keeled scrapers which are almost Aurignacian. In addition there were numerous fossil bones which belonged to mammals such as woolly rhinoceros (*Coelodonta antiquitatis*), reindeer (*Rangifer tarandus*), ibex (*Capra ibex*), hyena (*Crocuta crocuta*), marmot (*Arctomys marmotta*), wild horse (*Equus sp.*), bison (*Bison priscus*) and wolf (*Canis lupus*).

Dating In view of the 'cold weather' fauna and the tool culture the skeleton has been attributed to the Upper Pleistocene, probably the Fourth Glacial period (Würm or Last Glaciation).

Morphology The skeleton is almost complete and belonged to an adult male. It comprises the skull, twenty-one vertebrae, twenty ribs, one clavicle, two humeri, two radii, two ulnae, several hand bones including one scaphoid, one capitate, metacarpals I, II, III and V, as well as two proximal phalanges. The lower limbs are represented by two pelvic fragments, two femora, two patellae, parts of two tibiae, one fibula, one calcaneus, one talus, five metatarsals and several other fragmentary bones. The bones have been described by Boule (1911–1913) in detail, and the skeleton has been designated the type specimen of the species *Homo neanderthalensis*.

THE SKULL

The skull is large and well preserved, having a low vault and a receding forehead. The supra-orbital arches are large but the mastoid processes are small. The occipital bone protrudes into a characteristic bun-shape. The orbits are voluminous, the nose broad and the mandible stout but chinless. The bony palate is very large, but all the teeth are missing except for the upper and lower second left premolars.

THE POST-CRANIAL BONES

In general the limb bones are short and thick, with strong markings and large joints. The humeri are straight but the femora are bowed; similarly the radii are curved, having a medial concavity. The tibiae are short and stout, and their upper ends appear to be bent backwards into a position of retroversion. The fibula is robust. The hand and foot bones are not unlike those of modern man; the scaphoid and capitate seem small, but the metacarpals are stout with large heads. The foot bones are rugged, the talus in particular being high 29 and short-necked; the calcaneus is robust and has a prominent sustentacular shelf.

The vertebral column, as described by Boule (1911–1913), is said to have long, backwardly directed cervical spinous processes which are frequently bifid. The cervical and lumbar curvatures of the spine are obliterated and the remainder of the vertebrae are said to be short-bodied.

In view of the features of the post-cranial skeleton, Boule suggested that the stance of La Chapelle man (Neanderthal man) was stooping with flexed hips and knees and jutting head carriage, and his undoubtedly bipedal gait imperfect and slouching. Recent re-examination of the skeleton (Arambourg, 1955; Cave and Straus, 1957; Patte, 1955) has shown that there is evidence of gross deforming osteoarthritis present in the specimen and that Boule's reconstruction is faulty in a number of respects. In addition, comparison with other Neanderthal remains and a wider range of modern skeletal material has shown that many of the features recognized as being characteristically Neanderthal fall within the range of modern human skeletal variation. Whilst Cave and Straus (1957) do not deny the distinctive morphological characters of Neanderthal man, they suggest that he stood and walked as does modern man.

Dimensions   *Boule* (1911–1913), *Patte* (1955)
SKULL
Max. Length 208   Max. Breadth 156
Cranial Capacity 1,620 cc   Cranial Index 75·0
                              (Dolichocephalic)
Symphysial Angle of Mandible 104°
POST-CRANIAL BONES

	Length	Circumference		Torsion	
Right humerus	313		72	148°	
Left humerus	—		65	—	
Radius Length	235*	A/P diam.	12·0	Width	16·0
Ulna Length	255–260*	Upper Arm/Forearm			
		Carrying Angle 179°			
Capitate Length	24·0	Breadth	14·0		

* Restored.

	I	II	Metacarpals III IV		V
Length	44·5	73·0*	71·0*	—	54·0
Min. diam.	11·0	7·5	8·0	—	6·5

Femur Length 430* (R)   A/P diam. 31·0 (L)   Trans. diam. 29·0 (L)

Tibiae   —   Damaged   —

Talus Length/Breadth Index 107·5   Horiz. Angle of Neck 23°

Calcaneus Length 80*   Breadth 47
* Restored.

Numerous measurements and indices relating to the La Chapelle skeleton are quoted by Patte (1955).

Affinities The similarities between the skeleton from Neanderthal and that from La Chapelle-aux-Saints leave little doubt that they belong to the same species. Further finds have established that these men were widely distributed in Europe and the Near East during the Upper Pleistocene period. Their sudden disappearance and replacement by modern forms of man remains a topic for speculation and investigation. It is uncertain whether they became extinct because of the invasion of more advanced hominids, or became assimilated by the evolving population of modern man, or directly gave rise to modern man, an older view revived by Brace (1964). In a recent classification of the Hominidae, Campbell (1964) has identified Neanderthal man as a subspecies of *Homo sapiens* (*Homo sapiens neanderthalensis*).

Original Musée de l'Homme, Palais de Chaillot, Paris–16e, France.

Casts 1 Musée de l'Homme, Paris (Cranium and endocranium). 2 The University Museum, University of Pennsylvania, Philadelphia 4, Pennsylvania, U.S.A. (Restored skull only).

References BOULE, M. 1908
L'Homme fossile de La Chapelle-aux-Saints (Corrèze). *C. R. Acad. Sci. Paris. 147*, 1349-1352.
BOULE, M. 1908
L'Homme fossile de La Chapelle-aux-Saints. *Anthropologie 19*, 519-525.

The La Chapelle-aux-Saints Skeleton

BOUYSSONIE, A. and J., and BARDON, L. 1908
Découverte d'un squelette humain moustérien à La Chapelle-aux-Saints, Corrèze. *C. R. Acad. Sci. Paris 147*, 1414-1415.

BOULE, M. 1911-1913
L'Homme fossile de La Chapelle-aux-Saints. *Annls de Paléont. 6, 7, and 8.*

HRDLIČKA, A. 1930
The skeletal remains of early man. *Smithson. misc. Coll. 83*, 1-379.

ARAMBOURG, C. 1955
Sur l'attitude, en station verticale, des Néanderthaliens. *C. R. Acad. Sci. Paris 240*, 804-806.

PATTE, E. 1955
Les Néanderthaliens: Anatomie, Physiologie, Comparaisons. Paris; Masson et Cie.

CAVE, A. J. E., and STRAUS, W. L., Jnr. 1957
Pathology and posture of Neanderthal man. *Quart. Rev. Biol. 32*, 348-363.

BRACE, C. L. 1964
A consideration of hominid catastrophism. *Curr. Anthrop. 5*, 3-43.

CAMPBELL, B. 1964
Quantitative taxonomy and human evolution. In *Classification and human evolution*, 50-74. Ed. S. L. Washburn. London: Methuen and Co. Ltd.

OAKLEY, K. P., CAMPBELL, B. G. and MOLLESON, T. I. 1971
Catalogue of Fossil Hominids Part II: Europe. London: Trustees of the British Museum (Natural History).

The La Ferrassie Skeletons

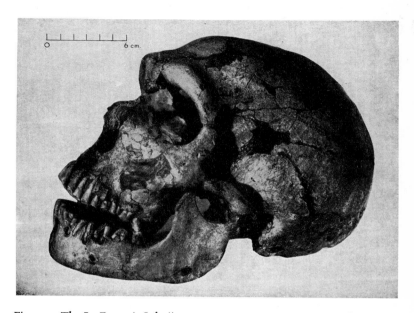

Fig. 10   The La Ferrassie I skull
Courtesy of the Director of the Musée de l'Homme

Synonyms and other names	*Homo neanderthalensis* (Boule, 1911–1913); *Homo sapiens neanderthalensis* (Campbell, 1964) Neanderthal man; Neandertal man; Ferrassie 1 to 6.
Site	La Ferrassie, north of Bugue, Dordogne, France.
Found by	R. Capitan and D. Peyrony, 17th September, 1909 (No. 1), 1910 (No. 2), 8th August, 1912 (Nos. 3, 4a and 4b), and in 1920 and 1921 (Nos. 5 and 6).
Geology	The remains were in a rock shelter. The stratigraphy, determined by a group of eminent French prehistorians, was said to be the same as that found at La Chapelle-aux-Saints. The skeletons were found below three Perigordian layers, four Aurignacian levels and one Chattelperonian layer at the bottom of a layer containing Mousterian tools; this in turn rested upon another layer containing Acheulean implements.

33

The bones of the first skeleton were lying in anatomical relation with each other and the layers were undisturbed.

Associated finds The tools found with the skeletons are of the Charentian Mousterian culture (Bourgon, 1957). The first mammalian bones recovered from the same deposit include those of mammoth (*Mammuthus primigenius*), hyena (*Crocuta crocuta*), pig (*Sus sp.*), ox (*Bos sp.*), red deer (*Cervus elaphus*) and horse (*Equus sp.*).

Dating The tools and fauna suggest an Upper Pleistocene date for the remains, probably during the Fourth Glacial period (Würm II) (Vandermeersch, 1965).

Morphology Originally only partly described (Boule, 1911–1913) the skeletons have now been reported more extensively (Heim, 1968, 1970 and 1974). La Ferrassie 1 is said to be the skeleton of an adult male and La Ferrassie 2 that of an adult female. La Ferrassie 3 is reported as the remains of a child of about 10 years of age, while number 4a is the remains of an 8-month foetus, 4b those of a neonate, La Ferrassie 5 the remains of a 6–7-month foetus and La Ferrassie 6 those of an infant of about three years. The adult skeletons are virtually complete while the immature skeletons are fragmentary.

THE SKULL

The male skull (No. 1) has stout supra-orbital ridges, a flattened vault with recession of the forehead, a protuberant occiput and small mastoid processes. The hard palate is broad and limited by an almost parabolic dental arcade. The mandible is perhaps more modern than that of La Chapelle man. It has a feeble chin, a deep mandibular notch and large coronoid processes. The teeth are all in place in both jaws, with no diastema. All of the teeth are badly worn so that no details of crown morphology are discernible.

In occlusion there is slight anterior displacement of the mandible so that the lower incisors project slightly in front of the upper incisors (negative overjet or underbite).

It has been suggested that the wear of the adult dentition should indicate that this hominid made use of the teeth as a 'tool', however, it is hard to distinguish this form of attrition from that caused by coarse particles in the diet (Wallace, 1975).   34

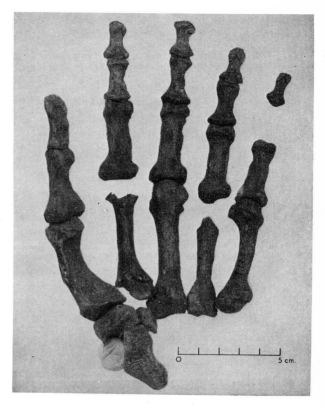

Fig. 11   The La Ferrassie I hand bones
An articulation of casts

THE POST-CRANIAL BONES

The limb bones resemble those of La Chapelle man in their principal features, tending to be robust with large joints. The tibiae have retroverted heads and the fibulae are stout.

The adult skeletons have well preserved feet and hands. Examination of the original foot bones (No. 1) shows that the great toe is not divergent, the longitudinal and transverse arches of the foot are well formed and that the general foot structure corresponds with that of modern man. These characteristics suggest that the stance and gait of La Ferrassie man differed little from that of *Homo sapiens*.

The La Ferrassie Skeletons

The hand bones are stout, the carpal arch is deep and the carpometacarpal joint of the thumb is saddle-shaped.

From a detailed metrical and morphological study of Neanderthal hand bones, including those from La Ferrassie, Musgrave (1970 and 1971) concluded that the Neanderthal hand shows unique features that may reflect environmental adaptations or the imperfect development of manual skill.

Dimensions   Heim (1974)
SKULL (La Ferrassie 1)
Max. Length 207·5   Max. Breadth 158
Cranial Capacity 1,681 cc.   Cranial Index 76·1
(Dolichocephalic)
Symphysial Angle of Mandible 85°
POST-CRANIAL BONES

	L.F. 1	L.F. 2
Right Humerus Max. Length	339	286
Left Humerus Max. Length	335	—
Right Tibia Max. Length)	c. 370	302
Left Tibia Max. Length)		300
Estimated Stature	1·7m	1·48m

Affinities   The features of the skeletons indicate that they belong to the group of Neanderthal men represented by the original Neanderthal skeleton and the skeleton from La Chapelle-aux-Saints. This group is characterized by numerous skeletal similarities as well as by their general contemporaneity, and had led to their designation as 'classic' Neanderthalers of the Würm glaciation. In a recent classification, Campbell (1964), Neanderthal man is identified as a subspecies of *Homo sapiens* (*Homo sapiens neanderthalensis*).

Originals   *Skeletons Nos. 1 to 4*—Musée de l'Homme, Palais de Chaillot, Paris—16e, France.
Skeletons Nos. 5 and 6—Musée des Eyzies, Les Eyzies, Dordogne, France.

Casts   Not available at present.

The La Ferrassie Skeletons

References CAPITAN, L. and PEYRONY, D. 1909
Deux squelettes humains au milieu de foyers de l'époque moustérienne. *Rev. anthrop. 19*, 402-409.

CAPITAN, L. and PEYRONY, D. 1911
Un nouveau squelette humain fossile. *Rev. anthrop. 21*, 148-150.

BOULE, M. 1911-1913
L'Homme fossile de La Chapelle-aux-Saints. *Annls de Paléont. 6*, 7
and 8.

CAPITAN, L. and PEYRONY, D. 1912
Station préhistorique de La Ferrassie. *Rev. anthrop. 22*, 76–99.

CAPITAN, L. and PEYRONY, D. 1912
Trois nouveaux squelettes humains fossiles. *Rev. anthrop. 22,* 439-442.

CAPITAN, L. and PEYRONY, D. 1921
Nouvelles fouilles à La Ferrassie, Dordogne. *C. R. Ass. franç. Av. Sci.
44e Session*, 540-542.

CAPITAN, L. and PEYRONY, D. 1921
Découverte d'un sixième squelette moustérien à La Ferrassie, Dordogne. *Rev. anthrop. 31*, 382-388.

HRDLIČKA, A. 1930
The skeletal remains of early man. *Smithson. misc. Coll. 83*, 1-379.

PATTE, E. 1955
Les Néanderthaliens: Anatomie, Physiologie, Comparaisons. Paris:
Masson et Cie.

BOURGON, M. 1957
Les industries moustériennes et prémoustériennes du Périgord. *Archs.
Inst. Paleont. Hum. 27*, 141.

CAMPBELL, B. 1964
Quantitative taxonomy and human evolution. In *Classification and
human evolution*, 50–74. Ed. S. L. Washburn. London: Methuen and
Co. Ltd.

VANDERMEERSCH, B. 1965
Position stratigraphique et chronologie relative des restes humains du
Paléolithique moyen dans le Sud-Ouest de la France. *Annls Paléont.
51*, 69–126.

HEIM, J-L. 1968
Les restes neandertaliens de La Ferrassie. Nouvelles données sur la
stratigraphie et inventaire des squelettes. *C. R. Acad. Sci., Paris. 266*,
576–578.

ANTHONY, J. and HEIM, J-L. 1970
La morphologie encéphalique de l'Homme de La Ferrassie I. *C. R.
Acad. Sci. Paris. 271*, 176–179.

HEIM, J-L. 1970
L'encéphale Néandertalien de l'Homme de La Ferrassie. *Anthropologie 74*, 527–572.

MUSGRAVE, J. H. 1970
An anatomical study of the hands of Pleistocene and Recent man.
Ph.D. thesis, University of Cambridge.

37

The La Ferrassie Skeletons

MUSGRAVE, J. H. 1971
How dextrous was Neanderthal man? *Nature 233*, 538–541.
HEIM, J-L. 1974
Les hommes fossiles de La Ferrassie (Dordogne) et le problème de la définition des Neandertaliens classique. *Anthropologie, 78*, 81–112 and 321–378.
WALLACE, J. A. 1975
Did La Ferrassie I use his teeth as a tool? *Curr. Anthrop. 16*, 393–401.

The Cro-Magnon Remains

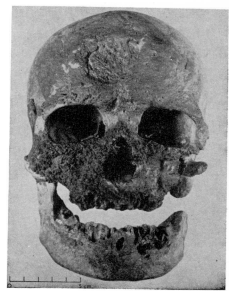

Fig. 12   The 'Old Man' from Cro-Magnon.
Courtesy of the Director of the Musée de l'Homme

Synonyms and other names	*Homo sapiens sapiens* Cro-magnon man; Le vieillard; The Old Man of Cro-Magnon; Cro-Magnon 1–5 (Oakley, Campbell and Molleson, 1971).
Site	Cro-Magnon, near Les Eyzies, Dordogne, France.
Found by	L. Lartet, 1868.
Geology	During the construction of a railway along the Vézère valley, excavation of a Cretaceous limestone cliff uncovered a rock shelter which had become filled by rock falls and debris. At the back of the shelter, under well recognizable occupation floors, were the remains of four adult skeletons, one neonate and some fragmentary bones. It appeared that the bones had been deliberately buried.
Associated finds	With the skeletons were numerous flint tools of Aurignacian manufacture, and large numbers of sea shells, some of which were pierced. The fossil bones of mammals recovered from the site included reindeer (*Rangifer tarandus*), bison (*Bison priscus*), mammoth (*Mammuthus primigenius*) and horse (*Equus sp.*).
Dating	The bones, unearthed by L. Lartet, an experienced geologist,

39

were stratigraphically contemporaneous with the deposit; the fauna and the tool culture suggested an Upper Pleistocene date for the burials, probably during the Fourth Glaciation (Würm or Last Glaciation).

Morphology The best-preserved skull belonged to an adult male known colloquially as the 'Old Man' of Cro-Magnon, although he was possibly a little under 50 years of age at his death (Vallois, 1937). It was upon the description of this skeleton that the principal features of the 'Cro-Magnon race' were based.

SKULL

The skull is virtually complete and undamaged except for missing mandibular condyles and a lack of teeth. The cranium is large and long with the vault somewhat flattened, and the outline of the skull from above is five-sided—the 'dolicho-pentagonal' form of some authors. In profile the forehead is steep and the superciliary ridges weak, the parieto-occipital region is curiously flattened in the midline and the parietal bosses are prominent. The face is broad and short with compressed rectangular orbits, whilst the tall narrow nasal opening and constricted nasal root make the nasal bones project. The subnasal maxilla slopes forward producing a moderate degree of alveolar prognathism. The bony palate is elevated centrally and grooved strongly for the palatal vessels and nerves. The dental arcade, as judged by the alveolar process of bone, is parabolic.

POST-CRANIAL BONES

The limb bones suggest that this individual was tall and muscular since the bones are long and the impressions for muscle and ligamentous attachments are strongly developed. The tibia is described as flattened and sabre-like (platycnemia), and the fibula deeply grooved.

Dimensions SKULL
Vallois and Billy (1965)
Max. Length 202   Max. Breadth 149·5
Cranial Capacity ± 1,600 cc   Cranial Index 74
(Dolichocephalic)

POST-CRANIAL BONES
Right Humerus Max. Length   no. 3 319
Left Humerus Max. Length   no. 3 320
Left Tibia Max. Length   no. 3 378

Affinities There is no doubt that the general skeletal morphology of this Cro-Magnon man shows that he belongs to *Homo sapiens*, but several features of the bones when taken as a group serve to distinguish this skeleton from those representing modern man. This group of characteristics has been used to define the 'Cro-Magnon race' of the Upper Pleistocene. Skeletal remains having these general features have been collected from many parts of western Europe and North Africa, including France (Abri-Pataud, Dordogne; Bruniquel; Chancelade; Combe-Capelle; Gourdan; Les Hoteaux dans l'Ain; La Madeleine; Mentone and Solutré), Germany (Obercassel; Stettin), Great Britain (Paviland, Glamorgan), Italy (Grimaldi), Czechoslovakia (Brno and Predmost, Moravia) and many other sites.

The remains from the Grottes des Enfants, Grimaldi, include two skeletons whose features have been regarded as negroid and thus may represent a distinct Upper Palaeolithic race; similarly the Chancelade remains have been likened to the skeletons of modern Eskimos and distinguished as the 'Chancelade race'.

Whether or not these are valid distinctions it is apparent that European Upper Palaeolithic men exhibited a wide range of variation in skeletal form. Moreover, it is clear that they used stone and bone in a variety of ways, producing implements, ornamental objects and cave paintings. The tools made by these peoples have been classified in many ways but three main groups predominate: the Aurignacian, Magdalenian and Solutrean cultures.

The origin of the Cro-Magnon people is still in some doubt because of the paucity of the fossil remains of their probable ancestors (for example Swanscombe and Fontéchevade); similarly their relationship with the Neanderthalers—at least some of whom were contemporaneous with Upper Pleistocene *Homo sapiens*—is an open question. It has been held variously that *Homo sapiens* displaced the Neanderthalers who became extinct or that the spreading *Homo sapiens* population absorbed the Neanderthalers or that *Homo sapiens* arose directly from a Neanderthal stock.

A new assessment of the Cro-Magnon remains has been given by Camps and Olivier (1970).

The Cro-Magnon Remains

Originals   Musée de l'Homme, Palais de Chaillot, Paris–16ᵉ, France.

Casts   1   Musée de l'Homme, Palais de Chaillot, Paris–16ᵉ, France. (Skull 1 and mandible, Skull 2 and Skull 3, one femur, two tibiae, one humerus, one ulna, one fibula).
2   The University Museum, University of Pennsylvania, Philadelphia 4, Pennsylvania, U.S.A. (Skull 1 reconstruction).

References   BROCA, P. 1865–1875
On the human skulls and bones found in the cave of Cro-Magnon, near Les Eyzies. In *Reliquiae aquitanicae*, 97-122. E. Lartet and H. Christy. Ed. T. R. Jones. London: Williams and Norgate.
BROCA, P. 1868
Sur les crânes et les ossements des Eyzies. *Bull. Soc. Anthrop. Paris 3*, 350-392.
PRUNER-BEY 1868
An account of the human bones found in the cave of Cro-Magnon in Dordogne. In *Reliquiae aquitanicae*, 73-92. E. Lartet and H. Christy. Ed. T. R. Jones. London: Williams and Norgate.
QUATREFAGES, A. DE and HAMY, E. 1874
La race de Cro-Magnon dans l'espace et dans le temps. *Bull. Soc. Anthrop. Paris 9*, 260-266.
QUATREFAGES, A. DE, and HAMY, E. 1882
Races humaines fossiles. In *Crania Ethnica*, 44-54 and 81-82. Paris: J. B. Baillière et Fils.
MORANT, G. 1930-1931
Studies of Palaeolithic man, IV. *Ann Eugen. Lond. 4*, 109-214.
BONIN, G. VON 1935
European races of the Upper Palaeolithic. *Hum. Biol. 7*, 196-221.
VALLOIS, H. V. 1937
La durée de la vie chez l'homme fossile. *Anthropologie 47*, 499-532,
VALLOIS, H. V., and BILLY, G. 1965
Nouvelles recherches sur les hommes fossiles de l'abri de Cro-Magnon. *Anthropologie 69*, 47-74.
DASTUGUE, J. 1967
Pathologie des hommes fossiles de l'abri de Cro-Magnon. *Anthropologie Paris 71*, 479-492.
CAMPS, G. and OLIVIER, G. (Eds) 1970 *L'homme de Cro-Magnon; Anthropologie et Archéologie*. Paris: C.R.A.P.E.
OAKLEY, K. P., CAMPBELL, B. G. and MOLLESON, T. I. 1971
Catalogue of Fossil Hominids Part II: Europe. London: Trustees of the British Museum (Natural History).

The Fontéchevade Skull Bones

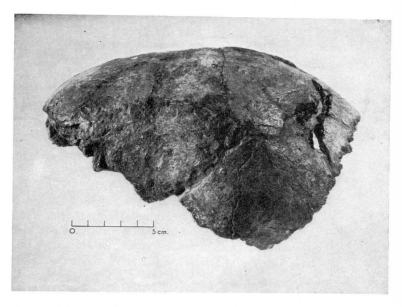

Fig. 13   The Fontéchevade II skull bones
Courtesy of the Director of the Musée de l'Homme

Synonyms *Homo* 'praesapiens'
and other names Praesapiens man (Vallois, 1949c); Fontéchevade I and II;
Fontéchevade 4 and 5 (Oakley, Campbell & Molleson, 1971);
Fontéchevade man.

Site Fontéchevade, near Montbron, Charente, about 17 miles east
of Angoulême, France.

Found by Mlle. G. Henri-Martin, 16th August, 1947.

Geology The Fontéchevade cave is cut into the wall of the valley of the
river Tardoire. During the excavation of the cave floor,
which had been previously investigated, a thick layer of
stalagmite was exposed. Beneath this layer were found deposits
which contained a fossil hominid calotte. The frontal bone
(Fontéchevade 1 or 4) was recovered from a block of breccia
in the laboratory and thus was not recognized *in situ* (Vallois,
1958). The layer of stalagmite has separated the cave filling

43

into two principal strata and indicates a considerable time difference between them.

Associated finds The original excavation had removed material containing Magdalenian, Aurignacian and Mousterian implements, but no hominid remains. The layer beneath the stalagmite contained primitive Tayacian flake tools in its upper part, whilst lower in this stratum the tools were Clactonian, an industry which is even more primitive in its workmanship.

The fossil fauna associated with the hominid remains includes rhinoceros (*Didermocerus sp.*), fallow deer (*Dama sp.*), tortoise (*Testudo graeca*) and bear (*Ursus sp.*): a warm/temperate interglacial fauna.

Dating The fossil layer has been attributed to the Third Interglacial period (Riss–Würm or Last Interglacial) in view of the fauna and the tool culture which it contains. The contemporaneity of the hominid fossils with the associated fauna has been suggested by means of the fluorine test (Oakley and Hoskins, 1951).

Morphology FONTÉCHEVADE II

The principal specimen consists of both parietals and part of the frontal bone. They are joined together and most of the sutures are obliterated. The right parietal is complete but the anterior portion of the frontal and the inferior border of the left parietal are missing. According to Vallois (1949b) the bones have been affected by fire. The external surfaces of the parietals are relatively featureless, having no parietal foramina, temporal lines or parietal bosses. The vault is somewhat flattened but the forehead is steep with no evidence of post-orbital constriction or supra-orbital torus. However Trinkaus (1973) has suggested that both post-orbital constriction and the presence of a supra-orbital torus are not excluded by the specimen. Viewed from above the skull is pentagonal, its greatest breadth being in the parietal region. The bones of the vault are absolutely and relatively thick.

FONTÉCHEVADE I

The smaller fragment is part of a frontal bone. It resembles the previous specimen in that it shows no evidence of a supra-orbital torus but its vault is thinner.

44

The Fontéchevade Skull Bones

Dimensions   *Vallois* (1958)   FONTÉCHEVADE II
Max. Breadth (154)   Length (195)
Cranial Index 78·9*   Cranial Capacity Approx. 1,470 cc
Biasterionic breadth (126)
( ) Specimen damaged
* Calculated figure

Affinities   The general features of the Fontéchevade calotte resemble
those of *Homo sapiens* (Vallois, 1949b), in particular the shape
of the forehead and the form of the parietal bones; on the
other hand it is distinguished by the flattening of the vault, the
thickness of the bones and the large biasterionic breadth. In
these latter features it can be likened to the Swanscombe skull
or to the Neanderthalers. The lack of a developed frontal
torus is considered by Vallois to rule out a close relationship
with the Neanderthalers or with pre-Neanderthal forms. He
states, therefore, that it is established that European *Homo
sapiens* is not derived from the Neanderthal men who preceded
him (Vallois, 1954). It was because of the similarities between
Swanscombe man and Fontéchevade man that they have
been grouped as 'Praesapiens' hominids until more is known
of the characters of the face and jaw of either or both forms
(Vallois, 1949b, 1954 and 1958). The term 'Praesapiens' is
alleged to be morphologically valid, but the extent to which
it indicates evolutionary continuity is not yet known. Vallois's
interpretation of the Fontéchevade remains has been questioned by Howell (1957), the evidence being regarded as 'not
fully conclusive'. Later, in a controversial article which has
been severely criticized, Brace (1964) casts doubt upon the
dating, reconstruction and interpretation of the Fontéchevade
material which he does not accept as representative of early or
pre-Neanderthal *Homo sapiens*. In a reconsideration of the
Fontéchevade fossils, Trinkaus (1973) concludes that the
frontal bone of Fontéchevade II (5) is too incomplete to
provide convincing evidence for the reconstruction of the
brow region and thus retains a cautious approach. But
Corruccini (1975), following a metrical analysis, states that in
parietal form it shows archaic features closest to those shown
by the Steinheim skull (*q.v.*).

45

The Fontéchevade Skull Bones

Originals Musée de l'Homme, Palais de Chaillot, Paris–16ᵉ, France.

Casts Not generally available at present.

References HENRI-MARTIN, G. 1947
L'Homme fossile tayacien de la grotte de Fontéchevade. *C. R. Acad. Sci. Paris 225*, 766–767
VALLOIS, H. V. 1949a.
L'Homme fossile de Fontéchevade *C. R. Acad. Sci. Paris 228*, 598–600
VALLOIS, H. V. 1949b
The Fontéchevade fossil man. *Am. J. Phys. Anthrop. 7*, 339–362.
VALLOIS, H. V. 1949c
L'origine de l'*Homo sapiens*. *C. R. Acad. Sci. Paris 228*, 949–951.
OAKLEY, K. P. and HOSKINS, C. R. 1951
Application du 'Test de la Fluorine' aux crânes de Fontéchevade. *Anthropologie 55*, 239–242.
HENRI-MARTIN, G. 1951
Remarques sur la stratigraphie de la grotte de Fontéchevade (Charente). *Anthropologie 55*, 242–247.
VALLOIS, H. V. 1954
Neandertals and Praesapiens. *J. R. Anthrop. Inst. 84*, 111–130.
HOWELL, F. C. 1957
The evolutionary significance of variation and varieties of 'Neanderthal man'. *Quart. Rev. Biol. 32*, 330–347.
VALLOIS, H. V. 1958
La grotte de Fontéchevade II. *Archs Inst. Paléont. hum. 29*, 1–262.
BRACE, C. L. 1964
A consideration of hominid catastrophism. *Curr. Anthrop. 5*, 3–43.
TRINKAUS, E. 1973
A reconsideration of the Fontéchevade fossils. *Am. J. Phys. Anthrop. 39*, 25–35.
CORRUCCINI, R. S. 1975
Metrical analysis of Fontéchevade II. *Am. J. Phys. Anthrop. 42*, 95–98.

The Arago Remains

Synonyms
and other names
Tautavel man; Arago man; Arago XXI (skull).

Site
Cave-site, Verdouble valley, at the southern tip of the Corbières mountain, near the village of Tautavel, 19 km. north-west of Perpignan, Pyrénées-Orientales.

Found by
H. de Lumley, 1964–1971.

Geology
The cave is in limestone and opens on to an escarpment about 50 m. above the Verdouble river where it emerges on to the Tautavel plain. The cave is about 35 m. deep and 10 m. broad. The Pleistocene deposits consist of several metres of sand and yellowish sandy loam of aeolian origin. These two layers are overlaid by a mineralized soil and then a thick stalagmitic layer that sealed in the earlier deposits; a breccia, in turn, overlies the stalagmitic layer.

Associated finds
Mammalian faunal remains recovered from the site include wolf (*Canis lupus*), panther (*Felis pardus*), cave bear (*Ursus spelaeus*), wild boar (*Sus scrofa*) wild ox (*Bos primigenius*) red deer (*Cervus elaphus*) reindeer (*Rangifer tarandus*), ibex (*Capra ibex*), rhinoceros (*Rhinoceros mercki*), horse (*Equus caballus cf. mosbachensis*), elephant (*Elephas sp.*) beaver (*Castor fiber*), rabbit (*Oryctolagus cuniculus cuniculus*). Other remains include those of tortoise (*Testudo sp.*) and some birds. (De Lumley, 1973). In addition to the faunal remains large assemblages of stone tools have been recovered from this site. Most of the tools have been attributed to the Tayacian with the exception of the upper levels which contained Middle Acheulean tools (de Lumley 1965 and 1971).

Dating
The dating of the hominid-bearing layers has been given as the early part of the Third Glaciation (Riss) on the basis of the fauna, particularly the rodents. (De Lumley 1971 and 1973; Chaline, 1971). No chemical or radiometric data is available concerning analyses that would give evidence of the relative or absolute age of the material.

Morphology
THE SKULL (ARAGO XXI)
The specimen was found on a living floor amidst numerous

47

fossil mammalian bones. It consists of a somewhat damaged and deformed face and partial vault. The frontal, sphenoid, zygomatic and maxillary bones are all present. Five teeth were in place, M^1–M^3 on the right side and M^1 and M^2 on the left side. The third molar shows little sign of wear. The supraorbital torus is very prominent and is separated from the vault of the skull by a deep supratoral sulcus. The frontal recedes and the vault is somewhat flattened. Seen from above the postorbital constriction seems marked but distortion of the frontal precludes a final judgement. The orbits are low and rectangular while the whole face seems to jut from the neurocranium, an appearance that is accentuated by the strong alveolar prognathism and flat anterior maxillary surface. So far no estimate of cranial capacity has been given but preliminary inspection reveals that it may be small relative to either modern or Neanderthal man.

The finds include a mandible with six teeth (Arago II), a half mandible with five teeth (Arago XIII), numerous isolated teeth, some phalanges, some parietal fragments and the anterior portion of an adult cranium. The mandibles are said to be massive in size but may represent male and female dimorphs. The bodies of the mandibles are moderately robust and well marked by digastric fossae on the inferior borders; the muscular markings for the mylohyoid and genial muscles are well developed. The rami are high and broad with shallow mandibular notches and large condyles. The larger of the two specimens (Arago XIII) has a receding symphysial region and no chin whereas the smaller (Arago II) still recedes but appears to possess an *incurvatio mandibulae* and a slight chin protuberance. The larger specimen has a prominent *planum alveolare*.

The teeth are said to differ in size in the two mandibles, the possible female (Arago II) having teeth of moderate size while those of the male (Arago XIII) are massive. The male mandible possesses right PM_1–M_2 and the female mandible right M_1 and M_2. No details of the cusp morphology have been given as yet.

Dimensions   Up to the present no measurements of the skull have been published.

MANDIBLES

	Arago II	Arago XIII
Length	108	124·5
Breadth	128	158
Symphysial angle	103°	106°

Affinities   Until such time as more data is available as to the dating and the morphology of the finds it is difficult to make an assessment of the affinities of this material. In general, however, such opinions that have been given seem to concur with that given by the finders. This view (de Lumley and de Lumley, 1973) is that the Arago finds are intermediate between *Homo erectus* and Neanderthal man in a number of respects. Mann and Trinkaus (1973), for example, state that Arago II (mandible) has its closest resemblance with the Heidelberg mandible (*q.v.*) and the Montmaurin mandible. Howells, however, takes the view, on the basis of the position of the third molar, that there are similarities to *Homo erectus* material from Ternifine (*q.v.*), Peking (*q.v.*), Java (*q.v.*) and several other sites. Since no dimensions are available and no metrical studies have yet been published these opinions must be regarded as provisional.

Original   Laboratoire de Géologie Historique, Faculté des Sciences, Place Victor Hugo, Marseille 3, France.

Casts   Not available at present.

References
DE LUMLEY, H. and DE LUMLEY, M-A. 1971
Découverte de restes humains anténéandertaliens datés du début de Riss à la Caune de l'Arago (Tautavel, Pyrénées–Orientales). *C.R. Acad. Sci. Paris 272*, 1729-1742.
CHALINE, J. 1971
L'âge des Hominiens de la Caune de l'Arago à Tautavel (Pyrénées-Orientales), d'après l'étude des Rongeurs. *C.R. Acad. Sci. Paris 272*, 1743-1746.
HOWELLS, W. W. 1971
Neanderthal man: facts and figures. *Proc. IX. Int. Cong. Anthrop. and Ethnol. Sci. Chicago; U.S.A. Yearb. phys. Anthrop.* 7-18.
DE LUMLEY, H. and DE LUMLEY, M-A. 1973
Pre-Neanderthal human remains from Arago cave in South-eastern France. *Yearb. phys. Anthrop. 17*, 162-168.
MANN, A. and TRINKAUS, E. 1973
Neanderthal and Neanderthal-like fossils from the Upper Pleistocene. *Yearb. phys. Anthrop. 17*, 169-193.

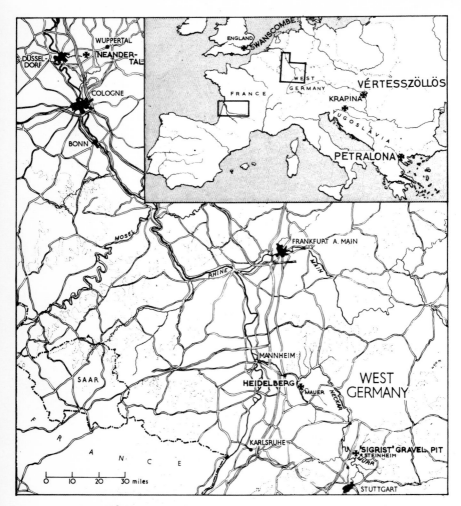

Fig. 14   Hominid fossil sites in Central Europe

The Neanderthal Skeleton

Synonyms and other names Homo neanderthalensis (King, 1864); Homo sapiens neanderthalensis (Campbell, 1964)
Neanderthal man; Neandertal man
Numerous other names have been proposed for Neanderthal forms, amongst them *Homo primigenius, Homo antiquus, Homo incipiens, Homo europaeus, Homo mousteriensis*. None of them precedes the name given by King and none is in common use.

Site The Neandertal valley, near Hochdal, about seven miles east of Dusseldorf, towards Wuppertal.

Found by Workmen; recognized by von Fuhlrott, August, 1856.

Geology The skeleton was uncovered when a cave deposit was disturbed during quarrying operations. The cave (Feldhofer grotto) was in Devonian limestone which formed the walls of the valley. Most of the valley no longer exists, neither does the cave. It is likely that much of the skeleton was lost before its significance was appreciated.

Associated finds Neither artefacts nor fossil mammalian bones were found with the remains.

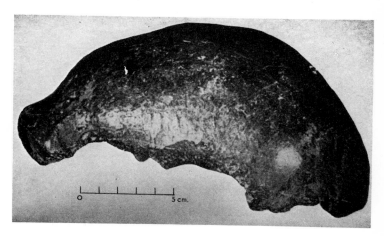

Fig. 15   The Neanderthal calotte
Courtesy of Professor J. S. Weiner

51

The Neanderthal Skeleton

Dating   This find was perhaps the earliest acceptable example of fossil man, but because of the circumstances of the find, the lack of stratigraphy or associated tools and the absence of a fossil fauna it was impossible to date the skeleton with any confidence. It is likely that the man from Neanderthal died during the Fourth Glaciation (Würm or Last Glaciation).

Morphology   The parts of the skeleton which have been preserved include the skull cap, one clavicle, one scapula, five ribs, two humeri, one radius, two ulnae, two femora and part of the pelvis.

THE CALOTTE

The vault of the skull is low, the supra-orbital ridges large and the occipital bone prominent. The sagittal suture and most of the coronal suture are obliterated, but the lambdoid suture is apparently still open. This degree of sutural fusion has led to the suggestion that the age at death was about 50 years.

THE POST-CRANIAL BONES

The limb bones are of rugged construction, having stout tuberosities and impressive muscular markings. The humeri are straight and cylindrical, but the radius is curved with an internal concavity. The radial tuberosity is very large, adumbrating a powerful biceps muscle. The left elbow was the site of an injury which severely limited movement at the joint. The femur is cylindrical and has the beginnings of a third trochanter.

Dimensions   Hrdlička (1930)

THE CALOTTE

Length 201   Breadth 147
Cranial Index 73·1 (Dolichocephalic)
Cranial Capacity 1,033 cc (Schaafhausen, 1858)
                  1,230 cc (Huxley, 1863)
                  1,234 cc (Schwalbe)

POST-CRANIAL BONES

	Length	*Mid-shaft diameter*	
Humerus	312	26	Schaafhausen, 1858
Femur	438	30	
Radius	239	15·5	

Affinities   The principal importance of this find is, perhaps, its historical interest, for in the years following *The Origin of Species* it caused intense controversy. Fuhlrott was willing to accept the antiquity of the bones and their contemporaneity with the mammoth, Virchow (1872) believed that the skull was pathological, whilst Blake (1864) considered that it belonged to an imbecile. Schaafhausen (1858) said that the bones were those of an ancient savage and barbarous race, but Huxley (1863) recognized the primitive features of the skeleton and could not accept that it was an intermediate form between man and apes; thus he proposed that Neanderthal man was an example of reversion towards a previous simian ancestor. King (1864) decided that the characters of the Neanderthal remains were so different from those of contemporary man that he proposed the name *Homo neanderthalensis*. The controversy continued unabated for many years until further finds convinced the sceptics that fossil man existed; of particular importance in this respect was the discovery of the La Chapelle-aux-Saints skeleton which shares numerous features with the Neanderthal remains. Boule was in no doubt that they were co-specific and declared the French find the type skeleton of the species; since then many Neanderthal skulls, skeletons and parts of skeletons have been found in Europe and the Near East as well as remains from Africa and Asia which are regarded by some anthropologists as being Neanderthaloid. Recently the suggestion has been revived that the Neanderthal morphology may be due to pathological change, one view impugning Vitamin D deficiency in youth (rickets) (Ivanhoe, 1970) and another syphilis (Wright, 1971). Neither suggestion has been adequately supported, or even justified, subsequently.

A general review of current theories relating to the origin of the Neanderthal group will be given later (p. 109).

Original Rheinisches Landesmuseum, Bonn, West Germany.

Casts 1. F. Krantz, Rheinisches Mineralienkontor, Bonn, West Germany. (Calotte and post-cranial bones.)
2. The University Museum, University of Pennsylvania, Philadelphia 4, Pennsylvania, U.S.A. (Calotte and endocranial cast.)

References SCHAAFHAUSEN, H. 1858
Zur Kenntnis der ältesten Rassenschädel. *Archiv. Anat. Phys. wiss. Medicin*, 453-478

FUHLROTT, C. VON 1859
Menschliche Uebereste aus einer Felsengrotte des Düsselthals. *Verh. naturk. Ver. der preuss. Rheinl. 16*, 131-153.

HUXLEY, T. H. 1863
Evidence as to man's place in nature. London: Williams and Norgate.

KING, W. 1864
The reputed fossil man of the Neanderthal. *Quart. J. Sci. 1*, 88-97.

BLAKE, C. C. 1864
On the alleged peculiar characters and assumed antiquity of the human cranium from the Neanderthal. *J. anthrop. Soc. Lond. 2*, 139-157.

VIRCHOW, R. 1872
Untersuchung des Neanderthal-Schädels. *Z. Ethn. 4*, 157-165.

HRDLIČKA, A. 1930
The skeletal remains of early man. *Smithson. Misc. Coll. 83*, 1-379.

CAMPBELL, B. 1964
Quantitative taxonomy and human evolution. In *Classification and human evolution*, 50-74. Ed. S. L. Washburn. London: Methuen and Co. Ltd.

IVANHOE, F. 1970
Was Virchow right about Neanderthal? *Nature 277*, 577-579.

WRIGHT, D. J. M. 1971
Syphilis and Neanderthal man. *Nature 229*, 409.

The Mauer Mandible

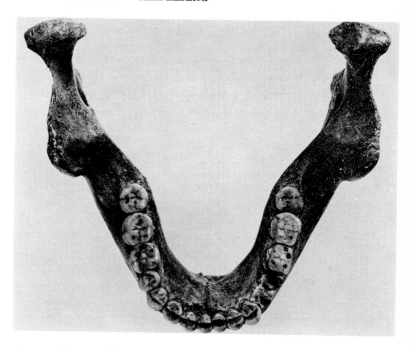

Fig. 16   The Heidelberg mandible. Occlusal view
*Courtesy of Dr R. Kraatz, Conservator, Geologisch-
Paläontologisches Institut der Universität, Heidelberg*

**Synonyms
and other names**   *Homo heidelbergensis* (Schoetensack, 1908); *Palaeanthropus
heidelbergensis* (Bonarelli, 1909a); *Homo (Euranthropus) heidel-
bergensis* (Arambourg, 1957)
Heidelberg man; Mauer man
Several other generic names have been proposed such as
Pseudhomo, Europanthropus, Rhenanthropus and *Maueranthropus.*
None has gained acceptance.

Site   The Rösch sandpit, half a mile north of the village of Mauer,
three miles south of Neckargemund and about six miles
south-east of Heidelberg, West Germany.

Found by   The mandible was found by a workman on the 21st October,
1907. It was shown to O. Schoetensack. (It is uncertain
whether the fossil was seen *in situ* by Schoetensack.)

55

The Mauer Mandible

Geology   The deposits exposed at the pit consist of a lower series of river sands, the Mauer sands, from the bed of the Neckar river. These sands are overlain by two layers of silt or loess which are approximately 50 feet thick and rest upon sandstone gravel. The mandible was found 80 feet from the surface, on a collapse about 3 feet from the base of the layer.

Associated finds   With the mandible was a well preserved fossil fauna of mammalian bones showing no evidence of water rolling. This fauna included Etruscan rhinoceros (*Didermocerus etruscus*), red deer (*Cervus elaphus*), straight-tusked elephant (*Palaeoloxodon antiquus*), bison (*Bison priscus*), moose (*Alces latifrons*), roe deer (*Capreolus capreolus*), horse (*Equus mosbachensis*), primitive bear (*Ursus sp.*) and beaver (*Castor fiber*). Less frequently there were carnivores including lion and primitive wolf.

No artefacts were found with the mandible.

Dating   The fauna is post-Villafranchian and pre-Second Interglacial (Mindel–Riss or Penultimate Interglacial), suggesting that the climate was probably warm/temperate. The stratigraphy and fauna indicates that the Mauer sands were laid most probably during the First Interglacial or possibly during an interstadial within the Second Glaciation. Chemical or radiometric dating of the find is unknown. Oakley (1964) confidently attributes Heidelberg man to the First Interglacial period.

Morphology   MANDIBLE

The mandible is large and stoutly built with broad ascending rami and rounded angles. The body is thick and particularly deep in the region of the premolars and molars; the mental symphysis is buttressed on its inner aspect but there is no simian shelf and no chin. The coronoid processes point forwards and laterally and the mandibular notches are shallow. The inner aspect of the bone shows well marked genial tubercles, and on each side a mylohyoid ridge above a shallow submandibular fossa. On the rami the openings of the inferior dental canals are overhung by lingular processes. On the left the groove for the mylohyoid nerve and vessels is particularly prominent.

Muscular markings include roughenings for the medial pterygoids, pits for the lateral pterygoids and marked inferior depressions for the digastric muscles. The mental foramina are multiple (Howell, 1960). With the exception of those for the masseters, the muscular impresses are in keeping with the size of the bone.

TEETH

The dental arcade is parabolic and the tooth row has no diastema. The permanent dentition is complete although the premolars and first two molars of the left side are broken. The teeth appear small and probably lie within the crown dimension ranges of modern man; however, only the molars are said to lie within the range of size variation of the Choukoutien pithecanthropines. The molar cusp pattern is dryopithecine. The molar teeth decrease in size in the order $M_2 > M_3 > M_1$, and all four are well worn. None of the molars has a cingulum or any sign of secondary enamel wrinkling. The premolars are bicuspid and the incisors somewhat swollen posteriorly, although not shovelled. The pulp cavities are moderately enlarged (taurodontism).

Molar Cusp Pattern (Howell, 1960)

	Lower Molars	
	Left	Right
M_1	Sub Y5	Sub Y5
M_2	+5	+5
M_3	+5	+5

Dimensions

	Boule and Vallois (1957)	Wust (1951)	Howell (1960)
Symphysial Height	—	37·0	34·0
Body Height (Behind M_1)	—	35·5 (M_2)	34·3
Body Thickness	23·0 (M_3)	19·5 (PM_1)	22·0 (M_1)
Ramus Height	66·0	69·0	71·0
Ramus Breadth	60·0	51·0	52·0
Mandibular Angle	—	—	105°
Total Length	—	126·0	120·0
Bicondylar Breadth	—	130·0	133·0

The Mauer Mandible

TEETH
Howell (1960)

		Lower Teeth (Crown Dimensions)				
		PM1	PM2	M1	M2	M3
Left	l	7·3	—	11·1	12·9	11·5
Side	b	—	—	—	—	11·3
Right	l	8·1	7·5	11·6	12·7	12·2
Side	b	9·0	9·2	11·2	12·0	10·9

These figures broadly correspond to those of Schoetensack (1908), differing principally in terminology and in Howell's reluctance to measure broken teeth.

Affinities This is an important, but isolated, mandible whose morphology shows a number of points of similarity with the mandibles of pithecanthropines (*Homo erectus*) of approximately equivalent Middle Pleistocene age. The skull to which it belongs is unknown.

For many years the relationships of the jaw have been discussed because of its combination of relatively advanced teeth in a mandible which has such archaic features as a receding chin and broad rami. Recently Howell (1960) has reappraised the Mauer mandible and, concluding that its dental and mandibular morphology shows 'fundamental differences' from those of Java and Peking man, he tentatively suggests that Mauer man belongs to the lineage represented later by the Montmaurin mandible and the early Neanderthalers. An alternative view places Mauer man within a larger group which could include the hominids from Java, Peking and Ternifine (Mayr, 1963). However, it has been suggested that the available material neither justifies the inclusion of Mauer man in any of the recently proposed subspecies of *Homo erectus*, nor warrants it being placed in a subspecific category of its own (Campbell, 1964).

Until the skull and/or the post-cranial bones are known, the affinities of this form must remain in doubt.

Original Geologisch-Paläontologisches Institut der Universität, Heidelberg, West Germany.

The Mauer Mandible

Casts 1. F. Krantz, Rheinisches Mineralien-Kontor, Bonn, West Germany.
2. The University Museum, University of Pennsylvania, Philadelphia 4, Pennsylvania, U.S.A.

References SCHOETENSACK, O. 1908
Der Unterkiefer des Homo heidelbergensis *aus den Sanden von Mauer bei Heidelberg*, 1-67. Leipzig: Wilhelm Englemann.
BONARELLI, G. 1909a
Palaeanthropus (n.g.) heidelbergensis (Schoet.) *Riv. ital. Palaeont. 15*, 26-31.
BONARELLI, G. 1909b
Le razze umane e le loro probabili affinita. *Boll. Soc. geog. ital. 10*, 827-851, 953-979.
WEINERT, H. 1937
Dem Unterkiefer von Mauer zur 30 jahrigen Wiederkehr seiner Entdeckung. *Z. Morph. Anthr. 37*, 102-113.
WUST, K. 1951
Üeber den Unterkiefer von Mauer (Heidelberg) im Vergleich zu anderen fossilen und mit besonderer Berucksichtigung der phyletischen Stellung des Heidelberger Fossils. *Z. Morph. Anthr. 42*, 1-112.
BOULE, M. and VALLOIS, H. V. 1957
Fossil men. London: Thames and Hudson.
ARAMBOURG, C. 1957
Les Pithécanthropiens, 33-41. In Mélanges Pittard. Brive-la-Gaillarde. Nizet.
HOWELL, F. C. 1960
European and northwest African Middle Pleistocene hominids. *Curr. Anthrop. 1*, 195-232.
MAYR, E. 1963
Animal species and their evolution. London: Oxford University Press.
CAMPBELL, B. 1964
Quantitative taxonomy and human evolution. In *Classification and human evolution* 50-74. Ed. S. L. Washburn. London: Methuen and Co. Ltd.
OAKLEY, K. P. 1964
Frameworks for dating fossil man. London: Weidenfeld and Nicolson.

The Steinheim Calvaria

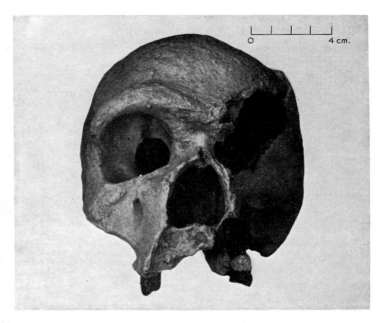

Fig. 17   The Steinheim skull (cast)
Frontal view

Synonyms   *Homo steinheimensis* (Berckhemer, 1936); *Homo* (*Protanthropus*)
steinheimensis (Berckhemer, 1937); *Homo cf. sapiens* (Le Gros
Clark, 1955); *Homo sapiens steinheimensis* (Campbell, 1964)
Steinheim man

Site   The Sigrist gravel pit, Steinheim on the river Murr, about
12 miles north of Stuttgart, Wurttemberg, West Germany.

Found by   Karl Sigrist, Jun., 24th July, 1933. Unearthed by F. Berckhemer.

Geology   The calvaria was found in Pleistocene gravels washed down
by the River Murr and deposited at Steinheim. The Sigrist pit
shows several distinct strata of sands and gravels overlain by a
layer of loess; these have been studied in detail by Berckhemer
(1925) and others who demonstrated that the gravels are of
two types and contain two separate faunal groups. This work
has been continued and amplified by Adam (1954a and b) who   60

suggests that four layers can be distinguished, each character-ized by its contained fauna, in particular by the elephant remains.

Associated finds No artefacts have been recovered from the Steinheim gravels. The layers have been named and characterized in the following way:

1. *Younger Mammoth Gravels*
Containing steppe-mammoth (*Mammuthus primigenius*) and woolly rhinoceros (*Coelodonta antiquitatis*).

2. *Main Mammoth Gravels*
Containing steppe-mammoth (*Mammuthus trogontherii-primigenius*), bison (*Bison priscus*), horse (*Equus steinheimensis*) and woolly rhinoceros (*Coelodonta antiquitatis*).

3. *Straight-tusked Elephant Gravels*
Containing straight-tusked elephant (*Palaeoloxodon antiquus*), bear (*Ursus arctos*), wild ox (*Bos primigenius*) and lion (*Felis sp.*). It was in this layer that Steinheim man was found.

4. *Older Mammoth Gravels*
Containing steppe-mammoth (*Mammuthus trogontherii*), horse (*Equus cf. mosbachensis*), rhinoceros (*Didermocerus kirch-bergensis*), bison (*Bison priscus*) and red deer (*Cervus elaphus*). According to Adam (1954a) the fauna in gravel 3 is clear evidence of an interglacial woodland environment at the time the deposit was laid down, for the straight-tusked forest elephant is almost confined to this layer whilst the steppe-mammoth occurs in each of the other three layers.

Dating In the past opinion has varied regarding the dating of this find, the gravels having been assigned by some to the Second (Mindel–Riss) and by others to the Third (Riss–Würm) Interglacials. In view of the evidence given above and other evidence, Howell (1960) states that there can be no question that the gravels in which the Steinheim calvarium was found are other than of Second Interglacial age (Great, Mindel–Riss or Penultimate Interglacial); this has been reaffirmed by Kurtén (1962) and Oakley (1964).

Morphology CALVARIA
A detailed description of the specimen has not yet been pub-lished, but several preliminary reports have appeared (Berck-hemer, 1933 *et seq.*). A longer account was published by

Weinert (1946). The specimen, which is distorted and damaged, consists of the cranial and facial skeleton of a young
adult. The left orbit, temporal and infra-temporal regions
and part of the left maxilla are missing, as is the pre-maxillary
region of both sides. The base of the skull is broken away
around the foramen magnum, but it is well preserved in its
anterior portion. The upper right second premolar tooth and
all of the molar teeth are in place.

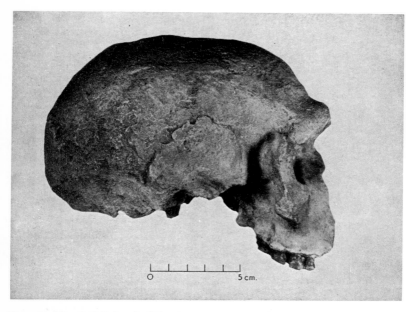

Fig. 18   The Steinheim skull (cast)
          Right lateral view

The vault of the skull is long, narrow and moderately flattened;
the supra-orbital torus is pronounced, the nasal opening wide
and the root of the nose depressed. In lateral view the degree
of facial prognathism is small, the mastoid processes small but
well defined and the occipital region well rounded. There is
no occipital chignon but a low torus extends to the asterion.
The greatest width of the cranium is its biparietal diameter.
TEETH
The premolar crown is symmetrical but has a large buccal   62

cusp and a smaller lingual cusp. The molars decrease in size in the order M1 > M2 > M3. The third molar is markedly smaller than the other two and all are moderately taurodont.

Dimensions	CRANIUM	*Weinert* (1936)	*Howell* (1960)
	Length	184	185
	Breadth	132–133	116
	Cranial Index	72·0	62·5
	Cranial Capacity	1,070 cc	1,150–1,175 cc

Affinities The precise relationships of this form are uncertain as the morphological features of the specimen have never been fully compared with those of other hominids. Nevertheless, there is little doubt that the Steinheim calvaria has few similarities with the known *Homo erectus* skulls of the Middle Pleistocene, but it does share some features with the Neanderthalers such as the heavy supra-orbital torus and the broad nasal opening. Perhaps its principal resemblances are with the Swanscombe remains (*q.v.*) which have been attributed to *Homo sapiens*; for example in such features as the high position of the maximum breadth, the contour of the occiput and the form and thickness of the bones of the vault. The teeth are small and sapient in form and molar size order, but primitive in that they show a moderate degree of taurodontism although this feature is not unknown in modern populations.

One view of the position of Steinheim man is that he represents an intermediate stage between *Pithecanthropus* (*Homo erectus*) and later varieties of *Homo*, thus being ancestral to both Neanderthal and modern man (Le Gros Clark, 1955). An alternative view (Weiner, 1958) does not accept Steinheim man as ancestral to the widely differing 'classic Neanderthaler' and modern man, but suggests that he represents a stage on a separate line leading to *Homo sapiens* and that Neanderthal and Rhodesian man have arisen from different and more primitive ancestors, for example Heidelberg, Montmaurin and Solo man.

It is likely that the evolutionary situation is more complicated than the earlier views would suggest, for it is possible that in the Middle Pleistocene a range of variable populations of early *Homo* were present, forms extending from pre-Neanderthal and pre-Rhodesian types to the more sapient groups   63

The Steinheim Calvaria

represented by Steinheim and Swanscombe man (Weiner, 1958; Napier and Weiner, 1962).

Original Institut für Anthropologie und Humangenetik der Universität, Tübingen 74, West Germany.

Casts   1. Staatliches Museum für Naturkunde, Stuttgart, West Germany. (Calvaria and reconstruction.)
2. The University Museum, University of Pennsylvania, Philadelphia 4, Pennsylvania, U.S.A. (Calvaria.)

References BERCKHEMER, F. 1925
Eine Riesenhirschstange aus den diluvialen Schottern von Steinheim a.d. Murr. *Jh. Ver. vater l. Naturk. Württemb. 81*, 99-108.

BERCKHEMER, F. 1933
Ein Menschen-Schädel aus den diluvialen Schottern von Steinheim a.d. Murr. *Anthrop. Anz. 10*, 318-321.

BERCKHEMER, F. 1936
Der Urmenschenschädel aus den zwischeneiszeitlichen Fluss—Schottern von Steinheim a.d. Murr. *Forsch. Fortsch. dtsch. Wiss. 12*, 349-350.

WEINERT, H. 1936
Der Urmenschenschädel von Steinheim. *Z. Morph. Anthr. 35*, 463-518.

BERCKHEMER, F. 1937
Bermerkungen zu H. Weinerts' Abhandlung 'Der Urmenschenschädel von Steinheim'. *Verh. Ges. phys. Anthrop. 2*, 49-58.

ADAM, K. D. 1954a
Die mittelpleistozänen Faunen von Steinheim an der Murr (Württemberg). *Quarternaria 1*, 131-144.

ADAM, K. D. 1954b
Die zeitliche Stellung der Urmenschen-Fundschicht von Steinheim an der Murr innerhalb des Pleistozäns. *Eiszeitalter Gegen. 4/5*, 18-21.

CLARK, W. E. LE GROS 1955
The fossil evidence for human evolution. Chicago: Chicago University Press.

WEINER, J. S. 1958
The pattern of evolutionary development of the genus *Homo. S. Afr. J. med. Sci. 23*, 111-120.

HOWELL, F. C. 1960
European and northwest African Middle Pleistocene hominids. *Curr. Anthrop. 1*, 195-232.

NAPIER, J. R., and WEINER, J. S. 1962
Olduvai Gorge and human origins. *Antiquity 36*, 41-47.

KURTÉN, B. 1962
The relative ages of the australopithecines of Transvaal and the

pithecanthropines of Java. In *Evolution und Hominisation*, 74–80. Ed.
G. Kurth. Stuttgart: Gustav Fischer Verlag.

CAMPBELL, B. 1964
Quantitative taxonomy and human evolution. In *Classification and human evolution*, 50–74. Ed. S. L. Washburn. London: Methuen and Co. Ltd.

OAKLEY, K. P. 1964
Frameworks for dating fossil man. London: Weidenfeld and Nicolson.

The Krapina Remains

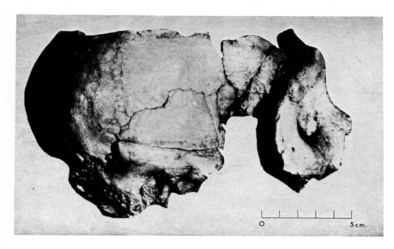

Fig. 19   The Krapina skull fragment
Courtesy of Professor J. S. Weiner

Synonyms and other names — *Homo primigenius var. krapinensis* (Gorjanovič-Kramberger, 1906) *Homo neandertalensis var. krapinensis* (Skerlj, 1953)

Site — Near the river Krapinica, a tributary of the Drave, which runs through the small town of Krapina, 25 miles north-west of Zagreb, Yugoslavia.

Found by — K. Gorjanovič-Kramberger, September, 1899–July, 1905.

Geology — The remains were uncovered during the excavation of a rock shelter which had become filled by fallen debris. The deposit consisted of river sands overlain by the layer containing the remains; this in turn was covered by sand derived from the Miocene sandstone and conglomerate walls of the shelter. There was some evidence of fire in the deposit.

Associated finds — It is believed that large numbers of stone implements were recovered from the site but relatively few have been described; they appear to be of Mousterian type and include points and scrapers. In addition Acheulean and 'pre-Aurignacian' tools have been mentioned (Skerlj, 1953).

66

Fossil mammalian bones found at the site included those of Merck's rhinoceros (*Didermocerus mercki*), cave bear (*Ursus spelaeus*), wild ox (*Bos primigenius*), beaver (*Castor fiber*), marmot (*Marmotta marmotta*) and red deer (*Cervus elaphus*).

Dating Originally the site was attributed to an early Interglacial date, First or Second Interglacial (Günz–Mindel or Mindel–Riss); a conclusion attributed to Gorjanovič–Kramberger by Skerlj (1953). However, other authors have regarded the site as being somewhat later, e.g. Third Interglacial (Riss–Würm) (Skerlj, 1953; Le Gros Clark, 1964); end of Third Interglacial and early stages of Würm I (Zeuner, 1940; Coon, 1962; Oakley, 1964); Göttweiger Interstadial of Würm Glaciation (Guenther, 1959).

Morphology The bones from Krapina are numerous but badly fragmented. They comprise the remains of at least 13 men, women and children; some of the bones show evidence of having been burned. Almost all the bones of the skeleton are represented, some several times, and many teeth.

SKULLS

Five skulls (A, B, C, D and E) are complete enough for study but only one (Skull C) gives any idea of the form of the cranium and face. This specimen consists of the upper part of the face together with part of the frontal bone, and the right temporal and parietal bones. It is probably the skull of an adolescent. The supra-orbital ridges—well marked laterally but divided in the midline—are separated by grooves from the somewhat retreating forehead. The bones of the vault are of modern form but the mastoid process is small. When reconstructed this skull was found to be brachycephalic.

The lower part of the Krapina face is represented by six maxillae (A–F) and nine mandibles (A–J, less I). These bones indicate that the palate was broad.

MANDIBLES

The most complete adult mandible (J) has a parabolic dental arcade bearing the permanent dentition, lacking only the first premolars and the left third molar. The body is relatively stout but there is no mandibular torus or simian shelf. The

67

condyles are flattened, although this is probably due to osteo-arthritis of the temporo-mandibular joint. There is no chin but the beginnings of a sub-mental trigone is discernible. In some specimens the mental foramina are large and single, in others small and multiple.

THE TEETH

The Upper Incisors are moderately shovelled and several have basal 'finger-like' tubercles.

The Canines resemble those of modern man but some have ridged margins.

The Premolars are bicusped, symmetrical and taurodont.

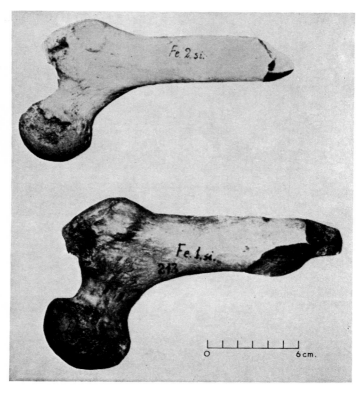

Fig. 20   The Krapina femoral heads
          Courtesy of Professor J. S. Weiner

68

The Krapina Remains

The Molars have four or five cusps, well marked anterior foveae and, in unworn teeth, secondary wrinkling of the enamel. An upper left first molar has a distinct Carabelli cusp and pit, whilst both upper and lower third molars show some reduction in size. Taurodontism is extreme in these molar teeth.

POST-CRANIAL BONES

Amongst the large collection are vertebrae, ribs, clavicles, scapulae, humeri, radii, ulnae, hand bones; also innominate bones, femora, patellae, tibiae, a fibula, a calcaneus, tali and other foot bones. Many of the bones are broken and incomplete.

In general there is little to distinguish these bones from their counterparts in modern man. They do not display the coarse modelling, heavy muscular markings and large joints typical of the 'classic Neanderthaler', e.g. La Ferrassie, La Chapelle aux-Saints.

Dimensions (Selected from *Gorjanović–Kramberger*, 1906)

SKULL C (Reconstructed)
Max. Length 178
Max. Breadth 149
Cranial Index 83·7 (Brachycephalic)

MANDIBLE J
Symphysial Height 42·3
Symphysial Thickness 15·0
Coronoid Height 73·0
Ramus Breadth (Middle) 37·0

TEETH

Lower Permanent Molars (Crown Dimensions)

Mandible J		M1	M2	M3
Left	l	11.0	12·5	—
Side	b	11·1	11·5	—
Right	l	11·4	11·5	11·6
Side	b	11·3	12·2	c 10·3

The Krapina Remains

POST-CRANIAL BONES

Clavicle (complete)	Length	149·5	(adult)
Clavicle (complete)	Length	59·4	(child)
Capitate	Length	27·5	
	Breadth	17·6	
Innominate bone (left)	Acetabular dia.	53·5	
Innominate bone (right)	Acetabular dia.	57·0	
Femora	Mean dia. of head I	52·7	
	Mean dia. of head II	44·3	
Talus	Length	53·3	
	Breadth	42·5	
	L/B Index	79·7	
Cuboid	Length	37·5	
	Breadth	27·4	

Numerous other measurements were given but many of them are of little value for comparative purposes since they are not standard dimensions.

Affinities The relationships of the Krapina people are difficult to assess because the features of their skeletons are mixed. The skulls are somewhat archaic in that the supra-orbital torus is pronounced, the mastoid processes small and the palate broad; similarly the mandible is chinless. The teeth also have primitive traits such as the extreme taurodontism and wrinkled enamel, but none have cingula. On the other hand the post-cranial bones are modern in their principal features.

Recent opinions on the dating have cast doubts upon the belief that these people antedate the 'classic Neanderthalers'. For convenience these hominids have been termed 'generalized Neanderthalers', although not regarded as specifically distinct from *Homo sapiens*; however, there may be grounds for giving this group subspecific rank (Le Gros Clark, 1964). Even this distinction is doubtful since modern schemes of hominid taxonomy only allow 'classic Neanderthal' man subspecific status within *Homo sapiens* (Campbell, 1964).

Originals National Museum of Geology and Palaeontology, Zagreb, Yugoslavia.

Casts National Museum of Geology and Palaeontology, Zagreb, Yugoslavia (Cranio-facial fragment, two mandibles, one femur).

References GORJANOVIĆ-KRAMBERGER, K. 1899
Vorläufige Mitteilung über den Krapinafund. *Mitt. anthrop. Ges. Wien 29*, 1, 65-68.

GORJANOVIĆ-KRAMBERGER, K. 1900
Der diluviale Mensch aus Krapina in Kroatien. *Mitt. anthrop. Ges. Wien 30*, 1, 203.

GORJANOVIĆ-KRAMBERGER, K. 1901-1905
Der paläolithische Mensch und seine Zeitgenossen aus dem Diluvium von Krapina in Kroatien. *Mitt. anthrop. Ges. Wien 31*, 164-197. *32*, 189-216. *34*, 187-199. *35*, 197-229.

GORJANOVIĆ-KRAMBERGER, K. 1906
Der diluviale Mensch von Krapina in Kroatia. Ein Beitrag zur Paläo-anthropologie. Wiesbaden: C. W. Kreidels Verlag.

ZEUNER, F. C. 1940
The age of Neanderthal man with notes on the Cotte de St. Brelade, Jersey, C.I. London: Occ. Paper No. 3 Univ. of London, Inst. of Archaeol.

SKERLJ, B. 1953
In *Catalogue des hommes fossiles*. Eds. H. V. Vallois and H. L. Movius, Jnr. *C. R. Cong. geol. Internat. Algiers 1952*. Section *5*, 250-251.

COON, C. S. 1962
The Origin of Races. London: Jonathan Cape.

CLARK, W. E. LE GROS 1964
The fossil evidence for human evolution. 2nd Ed. p. 76. Chicago: Chicago University Press.

GUENTHER, E. W. 1959
Zur Alters datierung der diluvialen Fundstelle von Krapina in Kroatien. In *Bericht über die 6 Tagung der Deutschen Gesellschaft für Anthropologie, Göttingen*, 202-209 (cited by Brace, 1964).

BRACE, C. L. (1964)
The fate of the 'Classic' Neanderthals: A consideration of hominid catastrophism. *Curr. Anthrop. 5, no. 1*, 3-43.

CAMPBELL, B. 1964
Quantitative taxonomy and human evolution. In *Classification and human evolution*, 50-74. Ed. S. L. Washburn. London: Methuen and Co. Ltd.

OAKLEY, K. P. 1964
Frameworks for dating fossil man, 303. London: Weidenfeld and Nicolson.

The Petralona Skull

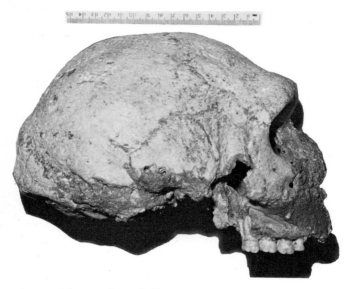

Fig. 21   The Petralona Skull
Right lateral view
Courtesy of Dr Christopher Stringer

Synonyms   Petralona 1
and other names

Site   Near Petralona, 37 km. south-east of Thessalonika, Khalki-dhiki, eastern Greece.

Found by   J. Malkotsis, J. Stathis, B. Avaramis, Chr. and Const. Sarijanides and St. Hantzaridés; 16th September, 1960.

Geology   Katsika Hill lies about 37 km. to the south-east of Thessalonika and consists of Mesozoic calcareous deposits. A stalagmitic cave was found accidentally in the hillside in May 1959, and later when the opening in the cave was enlarged and the cave explored fossil mammalian bones and teeth were found. Further exploration of the cave led to the discovery of the skull (Kokkoros and Kanellis, 1960). Further details of the geology and stratigraphy of the site are given in Poulianos (1971). Five principal strata are recognized (A–E from above downwards) containing eleven sub-strata made up of breccias,   72

The Petralona Skull

red earths, clays, and earths variously containing stalagmitic material, blackened fire-stones and ashes. Layers 1–10 (from above downwards) are said to show signs of human habitation.

Associated finds   Some stone tools and bone artefacts were recovered from the cave (Kanellis, 1964; Marinos, Yannoulis and Sotiradis, 1965; Poulianos, 1971). The tools have been ascribed to an early Mousterian culture and include quartz 'balls', scrapers and chopping tools made from imported materials. In addition some bone awls and bone scrapers were found. The more elaborate tools seem to come from higher in the sequence than those that are more simple in form.

The mammalian remains found in the Petralona cave deposits include those of cave bear (*Ursus spelaeus*), red deer (*Cervus elaphus*), and cave lion (*Leo spelaeus*) (Kokkoros and Kanellis, 1960). Later when the cave was investigated in greater detail further fossil mammals were identified including horse (*Equus*), fallow deer (*Dama dama*), giant deer (*Megaloceros*), wild ox (*Bos primigenius*) and wild goat (*Capra caucasia*) (Sickenberg, 1964). Further finds that have been reported include wolf (*Canis lupus*), hyena (*Crocuta crocuta*), and rhinoceros (*Didermocerus kirchbergensis*) (Poulianos, 1967).

Dating   The precise dating of the cave site seems to be a matter of some dispute within the general range of the Middle to Upper Pleistocene of Europe. The fauna has been said to indicate a date of either the end of the Third Interglacial (Riss–Würm) period or the beginning of the Last Glacial (Würm) period (Poulianos, 1967). In view of all the evidence it has been suggested later that a maximum date for the skull should be taken as 70,000 years B.P. (Poulianos, 1971). However, Sickenberg (1971) concluded that the Petralona fauna could well be pre-Second Glaciation (Mindel) in age.

Morphology   The skull was found 'suspended in the air (twenty-four centimetres above the floor) attached to a stalactitic column . . .' (Poulianos, 1971). In the same paper it is stated that the whole skeleton was found nearby lying on its right side in a contracted position. Unfortunately the postcranial skeleton was not preserved.

73

The Petralona Skull

The skull was covered with a calcareous incrustation that had served to preserve even the more fragile portions of the face. The only parts of the cranium that are missing are part of the right zygomatic bone and parts of both mastoid processes. The mandible has not been recovered. When the incrustation was removed from the cranial vault the shape of the skull emerged. It has large supraorbital ridges and a low vault with a retreating forehead. The occipital region is protuberant and leads down to a flattened nuchal plane. The face is large in most of its dimensions, but particularly so in the breadth of the upper face; the size of the mandible is estimated to have exceeded all known Neanderthal jaws in terms of its bicondylar breadth. Even the Heidelberg jaw (*q.v.*) is said to be too small to match the Petralona maxillae (Stringer, 1974). Further morphological details must await the cleaning of the skull and its full description.

Dimensions SKULL *Poulianos* (1971)
Max. Length 209   Max. Breadth 149
Cranial Index 71·6 (Dolichocephalic)
Cranial Capacity 1,220 cc

Affinities Until this skull is fully developed from its matrix and described it is obviously hard for those who have examined it to come to firm conclusions. Nonetheless preliminary opinions (Kokkoros and Kanellis, 1960) suggested that the skull is of Neanderthal type. Later the skull was also described as principally resembling the Neanderthals with particular emphasis on similarities with the Monte Circeo skull (Bostanci, 1964). Subsequent authors have varied widely in their views, some supporting the Neanderthal affinities (Jelinek, 1969; Brose and Wolpoff, 1971) others drawing attention to features that resemble those of the Rhodesian skull (*q.v.*) (Breitinger, 1964; Bostanci, 1964; Yrson, 1964), and one who emphasized its archaic features suggesting that it was an advanced *Homo erectus* (Hemmer, 1972). Poulianos (1974) finally suggested that the skull shows a mosaic of progressive and primitive features of 'Neanderthaloid' form and suggests a trend from Swanscombe–Steinheim–Vértesszöllös to Petralona.

The Petralona Skull

In a detailed multivariate metrical study of the Petralona skull Stringer (1974) concluded that it is cranially similar to the Neanderthalers but facially distinct from them. The closest resemblance is given to the Rhodesian skull and the Jebel Ighoud skull (*q.v.*) while its *Homo erectus* features were noted. Following a comparison of the Rhodesian and Petralona palates, Murrill (1975) felt that they both show a mixture of Pithecanthropic and Neanderthaloid dimensions. It seems clear that Petralona is not simply another 'classic' Neanderthaler but may represent a population of early forms comparable to the Rhodesian population from Africa and related to possible European equivalents such as Heidelberg and Vértesszöllös.

Original   Geological and Palaeontological Institute of the University of Thessalonika.

Casts   Not available at present.

References   ANON 1959
Newsletter *Makedonia:* Thessalonika (in Greek).
KOKKOROS, P. and KANELLIS, A. 1960
Découverte d'un crane d'homme paléolithique dans la peninsule Chalcidique. *Anthropologie 64*, 132-147.
BOSTANCI, E. 1964
An examination of the Neanderthal type fossil skull found in the Chalcidique peninsula. *Belleten Turk Tarih Kurumu Bosimeni 28*, 373-381.
BREITINGER, E. 1964
Report to the Moscow Anthropological Congress (quoted by Poulianos, 1967).
SICKENBERG, O. 1964
Die Saugetierfauna der Hohle Petralona bei Thessaloniki. *Geol. Geophys. Res. 9*, 1-16.
YRYSON, M. J. 1964
Report to the Moscow Anthropological Congress (quoted by Poulianos, 1967).
MARINOS, G., YANNOULIS, P. and SOTIRADIS, L. 1965
Palaeoanthropologische untersuchungen in der Hohle von Petralona-Chalkidi. (In Greek with a German summary.) *Wiss. Phys. Math. Fak. Univ. Thessaloniki*, 149-204.
POULIANOS, A. N. 1967
The place of Petralonian man among Palaeoanthropoi. *Anthropos (Brno) 19*, 216-221.

JELINEK, J. 1969
Neanderthal man and *Homo sapiens* in Central and Eastern Europe. *Curr. Anthrop. 10*, 475-503.
BROSE, D. S. and WOLPOFF, M. H. 1971
Early Upper Palaeolithic man and late Middle Palaeolithic tools. *Am. Anthrop. 73*, 1156-1194.
POULIANOS, A. N. 1971
Petralona: a Middle Pleistocene cave in Greece. *Archaeology 24*, 6-11.
HEMMER, H. 1972
Notes sur la position phylétique de l'homme de Petralona. *Anthropologie 76*, 155-162.
POULIANOS, A. N. 1974
The transitional period from the Neanderthaloid stage to *Homo sapiens*. (In Greek with English summary.) *Anthropos, 1*.
STRINGER, C. B. 1974
A multivariate study of the Petralona skull. *J. Hum. Evol. 3*, 397-404.
MURRILL, R. I. 1975
Z. Morph. Anthr. 66 (2), 176-187.

The Vértesszöllös Remains

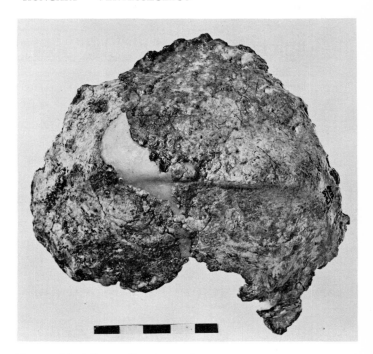

Fig. 22   The Vértesszöllös Occipital
External aspect
Courtesy of Dr Andor Thoma

Synonyms and other names	*Homo (erectus seu sapiens) palaeohungaricus* (Thoma, 1966); Vértesszöllös man
Site	Near the village of Vértesszöllös about 50 km. to the west of Budapest, Hungary.
Found by	Dr Laslo Vertes and his team from the Hungarian National Museum. Vértesszöllös I (excavated in 1964, recognized in 1965) and in 1965 on the 21st August, Vértesszöllös II.
Geology	The site lies at the foot of the Gerecse mountains in a quarry cut into the travertine deposits of the fourth terrace of the Danube system.
Associated finds	Four occupation layers were recognized in the deposits, each of dried mud that may have come from thermal springs. The

77

The Vértesszöllös Remains

lowest level (Level I) was 5 cm. thick and contains the human remains. Numerous artefacts were found, 'pebble-tools' and 'chopper-tools' as well as numerous flakes many of which were of small size (Buda Industry). Perhaps a 'microlithic variant of the industrial traditions of Afro-asia' (Kretzoi and Vertes, 1965).

The bones of fossil mammals that were recovered at the site included those of a beaver (*Trogontherium schmerlingi*), an etruscan rhinoceros (*Didermocerus etruscus*), a wild dog (*Canis etruscus*), a large sabre-toothed cat (*Epimachairodus sp.*) as well as a rich micro-mammalian fauna.

Dating The deposits underlie the Mindel loess and are within the fourth, or Mindelian terrace of the river Danube. The lower two occupation layers are said to correspond to a mild climate whereas the upper two layers correspond to a period of colder conditions. The presence of imprints of beech leaves from the lower layers suggests that the remains should be dated to a warm phase within the Second Glaciation (Mindel or Elster Glaciation). Traces of the use of fire were identified at the living site; thus Vértesszöllös man may rival Peking man as the earliest known fire-user. On the basis of a thorium/uranium estimation, Cherdyntsev, Kazachevsky and Kuzmina (1965) have given the date of the remains as approximately 350,000 years B.P.

Morphology VÉRTESSZÖLLÖS I
The first finds of human remains were several fragments of deciduous teeth from the lower dentition of a child. They include a left lower deciduous canine crown, a damaged deciduous second molar crown and two other tooth fragments (Thoma, 1967).

VÉRTESSZÖLLÖS II
The second find was an adult occipital bone broken into two fragments, an upper right portion and a lower left portion. The two parts fitted together leaving a small gap in the central region; the fragments were located by the internal vascular markings. The region of the lambda was slightly deformed

78

Fig. 23   The Vértesszöllös Occipital
Right lateral view
Courtesy of Dr Andor Thoma

by the downward and forward displacement of the sub-
lambdatic apex. In profile the bone can be divided into two
parts, an upper curved occipital portion and a lower flattened
nuchal portion, along the line of a well marked and undivided
occipital torus. The nuchal plane is incomplete and does not
include the foramen magnum whose borders are broken
away, nevertheless it was possible to reconstruct the position
of the opisthion with a reasonable degree of certainty.

The corners of the lambdoid sutures are well preserved and
correspond to the asterion of each side, however, there are
Wormian bones at lambda which render its exact location
dubious. The attachments of the sub-occipital muscles are   **79**

The Vértesszöllös Remains

well impressed on the nuchal plane, but the thickness of the bone bordering the foramen magnum does not constitute a postcondyloid tuberosity. There is a well marked occipitomastoid crest.

Internally the cerebellar fossae are small by comparison with the fossae for the occipital poles of the cerebral hemispheres, while the internal occipital protuberance lies well below the inion. The venous sinus impressions are distinct, the superior sagittal sinus passing directly into the right transverse sinus. Both transverse sinuses run directly on to the temporal bone at a point below each asterion without marking the parietal bones.

Dimensions Thoma (1966)
OCCIPITAL BONE
Biasterionic breadth 126·5; Lambda-opisthion chord 102; Lambda-inion chord 73; Inion-opisthion chord 56.
Cranial capacity estimated as in excess of 1,400 cc.

TEETH

Lower Teeth (Crown Dimensions)			
Vértesszöllös	I	dc_1	dm_2
	1	6·7	10·3
	b	5·4	6·4*

* damaged.

Affinities The remains have been studied by Thoma (1966, 1967 and 1969). In his view the occipital bone belonged to a male of under 30 years of age. In his description Thoma draws attention to the thickness and breadth of the bone, as well as the undivided occipital torus, as being primitive characters whereas the height and curvature of the upper segment of the bone are modern features. Similarly in his view the configuration of the brain is primitive while the capacity is large. However, the morphological comparisons and the metrical analyses taken together are said to indicate that while this man took his origin from *Homo erectus* he had differentiated from this group and thus occupies a phyletic position at the beginning of the progressive line represented by Swanscombe,

Fontéchevade and Quinzano (Thoma, 1966). Subsequently Stęślicka (1968) has stated that the occipital bone indicates Neanderthal affinities on the basis of its dimensions and proportions; on the other hand Wolpoff (1971 a and b) takes the view that the teeth and occipital bone together should be allocated to *Homo erectus.*

Originals   The Hungarian National Museum, Budapest, Hungary.

Casts   Not generally available at present.

References   CHERDYNTSEV, I., KAZACHEVSKY, V., and KUZMINA, E. A. 1965
Age of Pleistocene Carbonate Formation according to Thorium and Uranium Isotopes. *Geokhimiya 9,* 1085-1092.
KRETZOI, M., and VERTES, L. (1965)
Upper Biharian (Intermindel) Pebble-industry occupation site in western Hungary. *Curr. Anthrop. 6,* 74-87.
THOMA, A. 1966
L'occipital de l'homme Mindelien de Vértesszöllös. *Anthropologie 70,* 495-533.
THOMA, A. 1967
Human teeth from the Lower Palaeolithic of Hungary. *Z. Morph. Anthrop. 58,* 152-180.
STĘŚLICKA, W. 1968
W. sprawie Stanowiska Systematycznego Dolnoplejstocénskiej Kości Potylicznej Z Vértesszöllös *Przegąd Antropologiczny 2,* 267-274.
THOMA, A. 1969
Biometrische Studie über das Occipitale von Vértesszöllös. *Z. Morph. Anthr. 60,* 229-241.
THOMA, A. 1969
Le caractère aromorphotique de l'évolution humaine à la lumière des nouveaux fossiles. *Symp. Biol. Hung. 9,* 39-46.
WOLPOFF, M. H. 1971a
Is Vértesszöllös an occipital of European *Homo erectus? Nature 232,* 567-568.
WOLPOFF, M. H. 1971b
Vértesszöllös and the presapiens theory. *Am. J. Phys. Anthrop. 35,* 209-215.

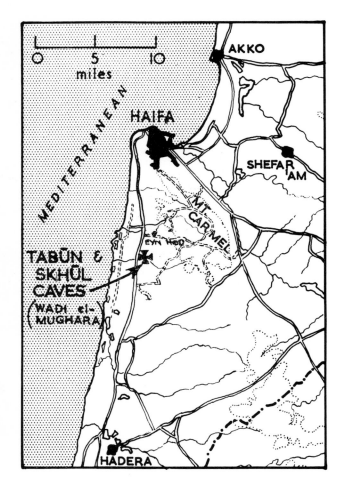

Fig. 24   The Tabūn and Skhūl sites

The Tabūn Remains

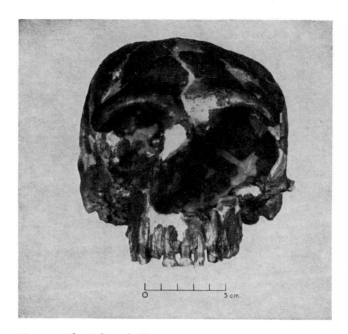

Fig. 25   The Tabūn skull
Frontal view
Photographed by courtesy of the Trustees of the British Museum (Nat. Hist.)

Synonyms and other names   *Palaeoanthropus palestinensis* (McCown and Keith, 1939); Tabūn I (C1) and II (C2); Mount Carmel man

Site   Mugharet et-Tabūn, Wadi el-Mughara, Mount Carmel, south-east of Haifa, Israel.

Found by   Joint Expedition of the British School of Archaeology in Jerusalem and the American School of Prehistoric Research, directed by D. A. E. Garrod (1929–1934).

Geology   The western slope of Mount Carmel overlooking the Wadi el-Mughara is penetrated by a number of caves which are hollowed into a steep limestone escarpment. Excavation of the floor of the Tabūn cave disclosed a number of archaeological

83

layers characterized by the flint tools and fossil mammalian bones which they contained. During the excavation a female skeleton and a male mandible were uncovered as well as some other fragmentary hominid bones.

Associated finds The occupation levels were lettered A–G from above downwards, and contained several types of implements.

Uppermost layer—A	Iron age and recent artefacts.
Upper layers—B, C and D	Levalloiso-Mousterian flake tools.
Lower layers—E and F	Upper Acheulean hand-axes, racloirs (scrapers), blades and retouched flakes (Yabrudian).
Lowermost layer—G	Tayacian tools, poor in quantity and quality.

The mammalian fossil fauna from layers C and D—the layers which contained the bulk of the hominid remains—was extensive and included hippopotamus (*Hippopotamus amphibius*), wild boar (*Sus gadarensis*), red deer (*Cervus elaphus*), fallow deer (*Dama dama mesopotamica*), wild ox (*Bos sp.*), gazelle (*Gazella sp.*), wild ass (*Equus hemionus*), hyena (*Crocuta crocuta*), rhinoceros (*Didermocerus cf. hemitoechus*) and roe deer (*Capreolus capreolus*). Above these layers, level B disclosed a distinct faunal break in that the fossil bones belonged to more modern forms, and the number of gazelle bones was much reduced whereas the fallow deer remains were increased. The new fauna did not contain hippopotamus, rhinoceros or other warm–climate species.

Dating Originally Garrod and Bate (1937) suggested that layers C and D belonged to the latter part of the Third Interglacial (Riss–Würm or Last Interglacial) and that the faunal break was evidence of the onset of the Würm Glaciation; however, carbon 14 dating of layer C at 45,000 ($\pm$2,000) years B.P. has altered this view. Recently (Garrod, 1962), has suggested that Tabūn C should be attributed to the second part of the Würm Glaciation and that the faunal break above this level was evidence of the onset of an interstadial period. However, Vogel and Waterbolk (1963) have given a radiocarbon date of 40,900 years B.P. $\pm$ 1,000 years for level C, the hominid layer for Tabūn I (C1) and Tabūn II (C2).

84

Morphology   Tabūn I (C1) is an almost complete adult female skeleton aged about 30 years. Tabūn II (C2) is an isolated mandible believed to be male. In addition there are five other specimens of post-cranial bones, at least some of which may belong to a further individual. The remains have been described in a lengthy monograph by Keith and McCown (1939).

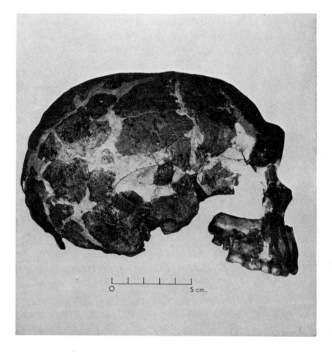

Fig. 26   The Tabūn I (C1) skull
Lateral view
Photographed by courtesy of the Trustees of the British Museum (Nat. Hist.)

TABŪN I

The skull was in fragments when found and needed extensive restoration. The cranium as restored is small and low-vaulted with a pronounced frontal torus, and the outline of the skull from above is shaped so that its maximum breadth is towards the back. There is no marked 'bun-formation', but an occipital

85

torus denotes powerful nuchal muscles. The mastoid process is small. The face is orthognathous and appears to be bent backwards to lie beneath the skull base.

The mandible has a stout body, broad rami and widely separated condyles. The symphysis is oblique and there is no chin. The mental foramina are double.

TABŪN II

The second individual is represented by a robust mandible which has a broad ramus and a better-developed chin.

TEETH

Tabūn I has a complete dentition, except for lack of the upper right third molar. The incisors are worn but lingual tubercles are present on both incisors and canines. The premolars are not remarkable other than they have some buccal swelling of the crown. The first two molars have four-cusped square crowns with well marked oblique ridges, but the third molars are triangular, lacking hypocones. A Carabelli pit is present on the first and second upper molars but it is absent on the third upper molars. The Tabūn II mandible has very worn teeth but the lingual tubercles are less developed than those of Tabūn I. All the teeth are present in the Tabūn II mandible except for the left lateral incisor, and in general the crown morphology of the lower teeth of Tabūn I and II is similar. The isolated teeth have corresponding characteristics, the molars having a dryopithecine cusp pattern.

THE POST-CRANIAL BONES

The long bones of the Tabūn I skeleton tend to be short, robust and somewhat bowed. In particular the radius and ulna are stout and curved. The vertebral column is modern in form but the bones of the shoulder are stout with strong muscular markings reminiscent of the bones of the 'classic Neanderthalers'. The pelvic girdle is low and narrow with flattened pubic rami. The tibia has a somewhat retroverted head, otherwise it is of modern form.

Dimensions   *McCown and Keith* (1939)

TABŪN I SKULL

Max. Length 183   Max. Breadth 141
Cranial Index 77·0 (Mesocephalic)
Cranial Capacity 1,271 cc (Pearson's formula)

The Tabūn Remains

TABŪN I AND II MANDIBLES	Tabūn I	Tabūn II
Symphysial Angle	61°	72°
Bicondylar Breadth	(133)	(130)
Length	(95)	(119)

TEETH

Tabūn I		I1	I2	C	PM1	PM2	M1	M2	M3
		Upper Teeth (Crown Dimensions)							
	l	9·0	7·3	7·9	7·5	6·5	10·8	10·5	8·3
	b	8·2	7·7	8·8	9·8	9·6	11·5	11·7	10·2
Side		R	L	R	R	R	L	R	L

Tabūn I		I1	I2	C	PM1	PM2	M1	M2	M3
		Lower Teeth (Crown Dimensions)							
	l	5·7	6·7	8·0	7·0	5·9	10·0	11·2	10·9
	b	7·0	7·6	8·3	8·5	8·7	10·5	10·6	9·8
Side		R	R	R	R	R	R	R	L

Tabūn II		I1	I2	C	PM1	PM2	M1	M2	M3
		Lower Teeth (Crown Dimensions)							
Right	l	5·9	6·1	8·0	7·8	7·9	11·0	10·8	11·5
Side	b	8·0	8·2	9·0	9·0	9·5	11·0	11·0	10·8

TABŪN I POST-CRANIAL BONES

Length of Humerus	287	Left
Length of Radius	222	Left
Length of Ulna	243	Left
Length of Femur	(416)	Right
Length of Tibia	310	Left

( ) Approximate measurement

Affinities The Tabūn and Skhūl remains were found in separate caves literally within yards of each other. Both groups of bones were described by McCown and Keith (1939) in the same monograph. It is convenient to discuss their relationships together, despite the recent suggestion that chronologically they may be separated by 10,000 years (Higgs, 1961).

Originals *Tabūn I* and some isolated specimens from Layer C: British Museum (Natural History), Cromwell Road, South Kensington, London, S.W.7.

The Tabūn Remains

Tabūn II and some other fragments: Museum of the Department of Antiquities, Jerusalem.

The remaining specimens: Peabody Museum, Harvard University, Cambridge, Mass., U.S.A.

Casts Tabūn II, mandible: The University Museum, University of Pennsylvania, Philadelphia 4, Pennsylvania, U.S.A.

References GARROD, D. A. E., and BATE, D. M. A. 1937
The stone age of Mount Carmel. Vol. 1 Excavations at the Wady el-Mughara. Oxford: The Clarendon Press.
McCOWN, T. D., and KEITH, A. 1939
The stone age of Mount Carmel. Vol. 2 The fossil human remains from the Levalloiso-Mousterian. Oxford: The Clarendon Press.
HIGGS, E. S. 1961
Some Pleistocene faunas of the Mediterranean coastal areas. *Proc. prehist. Soc. 27*, 144-154.
BROTHWELL, D. R. 1961
The people of Mount Carmel. *Proc. prehist. Soc. 27*, 155-159
HIGGS, E. S., and BROTHWELL, D. R. 1961
North Africa and Mount Carmel: Recent developments. *Man 61*, 138-139.
GARROD, D. A. E. 1962
The Middle Palaeolithic of the Near East and the problem of Mount Carmel man. *J. R. Anthrop. Inst. 92*, 232-259.
VOGEL, J. C. and WATERBOLK, H. T. 1963
Gröningen radiocarbon dates. *Radiocarbon 5*, 172.

The Skhūl Remains

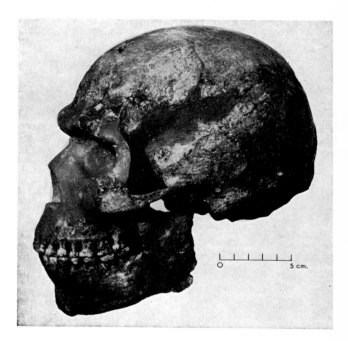

Fig. 27   The Skhūl V skull
Reconstructed by C. E. Snow
Left lateral view
Courtesy of the Trustees of the British Museum (Nat. Hist.)

Synonyms **Palaeoanthropus palestinensis** (McCown and Keith, 1939)
and other names Mount Carmel man

Site Mugharet es-Skhūl, Wadi el-Mughara, Mount Carmel, south-east of Haifa, Israel.

Found by Joint expedition of the British School of Archaeology in Jerusalem and the American School of Prehistoric Research, directed by D. A. E. Garrod (1929–1934).

Geology The western slope of Mount Carmel overlooking the Wadi el-Mughara is penetrated by a number of caves which are hollowed into a steep limestone escarpment. The smallest cave, really a rock shelter, is the Mugharet es-Skhūl. Excavation

89

of the limestone breccia which formed the cave floor uncovered several archaeological layers containing flint tools and fossil mammalian bones. Amongst these remains there were a number of hominid bones, belonging to at least ten individuals who appeared to have been intentionally buried. Most of the bones were in proper relation with each other showing little sign of disturbance, but no grave furniture or grave outline was found. The only object directly related to the skeletons was a wild boar mandible in the arms of Skhūl V. Stratigraphically the burials were all contemporaneous with the deposit.

Associated finds The flint implements recovered from the cave all belong to the Levalloiso-Mousterian culture. No Acheulean tools precede them as in the Tabūn cave. The principal finds were made in layers B1 and B2 and consist of Mousterian points, racloirs (scrapers), Levallois flakes, cores and burins.

The fossil mammalian fauna included wild ox (*Bos sp.*), hyena (*Crocuta sp.*), hippopotamus (*Hippopotamus amphibius*), rhinoceros (*Didermocerus cf. hemitoechus*), wild ass (*Equus hemionus*) gazelle (*Gazella sp.*), fallow deer (*Dama dama mesopotamica*), roe deer (*Capreolus capreolus*), red deer (*Cervus elaphus*), boar (*Sus gadarensis*) and small carnivores (*Felis sp.*). The wild ox was the commonest species in the assemblage.

Dating The fauna of the Skhūl cave deposit can be correlated with the fauna of Tabūn layer C, prior to the faunal break of layer B. The tools found at the Tabūn and the Skhūl skeleton levels are almost indistinguishable in their workmanship; in view of this Garrod and Bate (1937) believed that the two sites were contemporaneous and could be attributed to the end of the Third Interglacial (Riss–Würm or Last Interglacial). By correlating the faunal changes found at several Mediterranean coastal sites, Higgs (1961) produced evidence which led him to suggest that the Skhūl site may be as much as 10,000 years more recent than the Tabūn site, and thus within the Gottweiger Interstadial (Higgs and Brothwell, 1961). This view was rejected by Garrod (1962) who maintained that the Tabūn and Skhūl remains are broadly contemporaneous, and suggested a tentative radiocarbon date of 45,000 ($\pm$2,000) years B.P. for layer C of Tabūn, thus within the second part of the Würm Glaciation.

The Skhūl Remains

Morphology The remains uncovered in the Skhūl cave have been identified by McCown and Keith (1939) as follows:

Designation	Status	Sex	Age in years	Bones
Skhūl I	Infant	♂?	4–4½	Skeleton
Skhūl II	Adult	♀	30–40	Skeleton
Skhūl III	Adult	♂	—	Left leg bones
Skhūl IV	Adult	♂	40–50	Skeleton
Skhūl V	Adult	♂	30–40	Skeleton
Skhūl VI	Adult	♂	30–35	Skeleton
Skhūl VII	Adult	♀	35–40	Skeleton
Skhūl VIII	Child	♂?	8–10	Lower limb bones
Skhūl IX	Adult	♂	Approx. 50	Incomplete skeleton
Skhūl X	Infant	♂?	5–5½	Mandible and humeral fragment only

In addition there were sixteen isolated specimens.

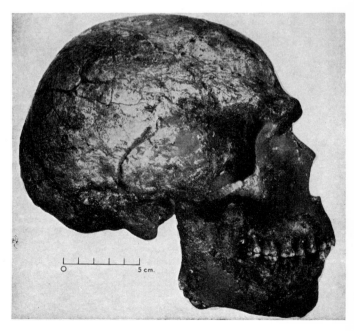

Fig. 28   The Skhūl V skull
Reconstructed by C. E. Snow
Right lateral view
Courtesy of the Trustees of the British Museum (Nat. Hist.)

The Skhūl Remains

The best-preserved skull is that which belongs to Skhūl V. The cranial vault of this specimen is high, the supra-orbital torus marked and the occipital region full and rounded. The facial skeleton is somewhat prognathic and meets the vault at the depressed nasal root. The bony palate is broad but the temporomandibular joint and the mastoid process are of modern form. The external auditory meatus is tall but the greater wing of the sphenoid and the orbital process of the zygomatic bone have some archaic features. The angle of the cranial base (basispheniod/basioccipital angle) is modern, as is the plane of the foramen magnum.

MANDIBLE
In general the Skhūl mandibles have chins although this feature is poorly marked in Skhūl V. The mental foramina are single and the mandibular condyles do not differ appreciably from those of modern man, but some of the specimens have traits which recall the Tabūn mandibles.

TEETH
The teeth of the Skhūl specimens are worn but do not differ very much from the teeth of the Tabūn skull and mandible. Individual variation is discernible in that Skhūl VII has teeth with a few archaic features whereas Skhūl IV and V have more modern teeth. Skhūl V has bony evidence of dental sepsis in the form of several apical abscess cavities and the bony changes of pyorrhoea, but no dental caries.

POST-CRANIAL BONES
The post-cranial bones have mixed characteristics, the majority being very like those of modern man but others resembling those of the Tabūn skeleton. The Skhūl long bones are long and slender contrasting with the stout, curved and big-jointed bones of the so-called 'classic Neanderthaler'. The hands are large and broad with well developed thumbs. Similarly the feet are large, stoutly constructed and well arched, without doubt feet well adapted to a propulsive bipedal gait. The axial skeleton has few distinctive features but the fourth segment of the sternum of Skhūl IV is long and resembles that of the Tabūn skeleton. The ribs are variable in form, Skhūl V ribs having a thicker and more rounded cross-section than those of Skhūl IV which are flattened and modern.

Dimensions McCown and Keith (1939)

SKULLS (Unrestored)	Skhūl IV	Skhūl V
Max. Length	(206)	192
Max. Breadth	(148)	143
Cranial Index	71·8	74·5
Cranial Capacity (Pearson's formula)	1,554 cc	1,518 cc

MANDIBLES

Bicondylar Width	(132)	(133)
Length	(118)	109
Symphysial Angle	75°	69°

POST-CRANIAL BONES	Skhūl IV	Skhūl V
Length of Humerus	—	Left 379
Length of Ulna	—	Right 270
Length of Radius	—	Right 236
Length of Femur	—	Right and Left (518)
Length of Tibia	—	Left (412)

TEETH

				Upper Teeth (Crown Dimensions)				
Skhūl V	I_1	I_2	C	PM_1	PM_2	M_1	M_2	M_3
Right l	8·5	7·0	8·7	8·2	7·0	11·0	10·8	9·1
Side b	7·5	7·2	9·5	10·8	10·8	12·5	12·2	11·8

				Lower Teeth (Crown Dimensions)				
Skhūl V	I_1	I_2	C	PM_1	PM_2	M_1	M_2	M_3
l	5·0	6·4	8·0	8·2	7·7	11·3	11·6	11·4
b	6·4	7·0	9·0	9·2	9·1	11·5	11·4	10·5
Side	L	R	R	L	R	R	R	R

( ) Approximate measurement

SKHŪL V (Restored by C. E. Snow, 1953)

SKULL	Before restoration	After restoration
Length	192	192
Breadth	143	144
Cranial Index	74·5	75·0
Cranial Capacity (Water displacement of cavity cast)	1,450 cc	—
(Pearson's formula)	1,518 cc	1,518 cc

MANDIBLE	*Before restoration*	*After restoration*
Length	109	107
Bicondylar Width	(132)	131
POST-CRANIAL BONES		
Humerus (Left)	379	378
Ulna (Right)	270	(280)
Radius (Right)	236	(254)
Femur (Right)	518	(505)
Tibia (Left)	412	(438)

( ) Approximate measurement

Affinities At first McCown and Keith (1939) admitted that they believed the Tabūn and Skhūl remains to belong to two distinct peoples, but when their studies were completed they concluded that the burials represented a single population of one species or race, for three particular reasons. In the first place the dental characters of the two groups were uniform; secondly, the associated tool cultures were nearly identical; and, lastly, according to Garrod and Bate (1937) they were found at the same locality and were contemporaneous. None the less, anatomically the Skhūl skeletons were considered to be of a later type, representing one extreme of a variable series. McCown and Keith rejected the possibility of hybridity between Neanderthal and Cro-Magnon man as an explanation of the mixed characteristics of the Mount Carmel skeletons on the grounds that there were no known previous fossils of either of these groups in the vicinity; they suggested that the nearest relatives of Mount Carmel man were the Krapina people (*q.v.*) of approximately the same date. Subsequent authors were divided between 'extreme variability' and 'hybridization' as possible explanations for the morphological findings until Higgs (1961) cast doubt upon the dating of the Skhūl remains. The anthropological implications of the new dating were discussed by Brothwell (1961). He concluded that broadly the two groups of remains can be separated into an earlier Palestinian Neanderthal population (Tabūn) who were replaced by a more advanced sapient people (Skhūl), a concept supported by the evidence of the Djebel Kafzeh remains which are similar to those from Skhūl, and the remains from Gallilee and Shanidar which resemble the Tabūn specimen. 94

The Skhūl Remains

Subsequent comments by a number of authors do not alter this opinion to any great extent.

Originals *Skhūl I and IV*: Museum of the Department of Antiquities, Jerusalem.
Skhūl IX: British Museum (Natural History), Cromwell Road, South Kensington, London, S.W. 7.
Skhūl II, III, V, VI, VII, VIII and isolated teeth: Peabody Museum, Harvard University, Cambridge, Mass., U.S.A.

Casts *Skhūl I, IV and V, also Skhūl V* (C. E. Snow reconstruction): The University Museum, University of Pennsylvania, Philadelphia 4, Pennsylvania, U.S.A.

References GARROD, D. A. E., and BATE, D. M. A. 1937
The stone age of Mount Carmel. Vol. I Excavations at the Wady el-Mughara. Oxford: The Clarendon Press.
McCOWN, T. D., and KEITH, A. 1939
The stone age of Mount Carmel. Vol. II The fossil human remains from the Levalloiso-Mousterian. Oxford: The Clarendon Press.
SNOW, C. E. 1953
The Ancient Palestinian Skhūl V reconstruction. *Amer. Sch. prehist. Res. Bull. 17*, 5-12.
HIGGS, E. S. 1961
Some Pleistocene faunas of the Mediterranean coastal areas. *Proc. prehist. Soc. 27*, 144-154.
BROTHWELL, D. R. 1961
The people of Mount Carmel. *Proc. prehist. Soc. 27*, 155-159.
HIGGS, E. S., and BROTHWELL, D. R. 1961
North Africa and Mount Carmel: Recent developments. *Man 61*, 138-139.
GARROD, D. A. E. 1962
The Middle Palaeolithic of the Near East and the problem of Mount Carmel man. *J. R. Anthrop. Inst. 92*, 232-259.

The Jebel Kafzeh Remains

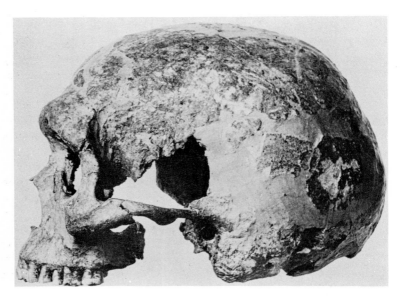

Fig. 29   The Jebel Kafzeh VI Skull
Left lateral view
Courtesy of Dr Bernard Vandermeersch

*Synonyms
and other names*
Homo sapiens sapiens (Vandermeersch, 1971); Jebel Qafza man; Djebel Qafzeh man

Site
A cave site 2·5 km. from Nazareth, towards the south, on the south-west flank of Mount Kafzeh, Israel.

Found by
Kafzeh 1 and 2, R. Neuville, 1933; Kafzeh 3–7, R. Neuville and M. Stekelis, 1934 (Kafzeh 3–6) and 1935 (Kafzeh 7); Kafzeh 8–11, B. Vandermeersch, 1966 (Kafzeh 8), 1969 (Kafzeh 9 and 10), 1970 (Kafzeh 11).

Geology
The Jebel Kafzeh cave is extremely large, about 20 m. broad and 12 m. deep. It communicates with the exterior through a big 'vestibule' which gives place to a slope. In all, 24 layers have been described including breccias, stalagmitic layers and

96

limestone layers of differing colours, that have been numbered I–XXIV from above downwards. Layers I–III correspond to breccias against the rock wall and layers IV–VII are relatively recent. Below these layers are a series of occupation levels, IX–XVIII, containing stone tools and hearths. Bed XVII, the layer containing Kafzeh 8–10, is split into two by a layer of yellow gravel. Layer XVII of Vandermeersch (1966) corresponds with Layer L of Neuville in its upper portion and the yellow gravel corresponds with Layer M of Neuville. Kafzeh 3, 6 and 7 are believed to come from the same layer (XVII) from the examination of the site records kept by Neuville (Vandermeersch, 1971). Kafzeh 4 and 5 are less certain in their provenance since they were recovered from a trial excavation in front of the vestibule. With this reservation it seems likely that they also derive from Layer XVII. Kafzeh 8, 9 and 10 were burials in the deposits of the vestibule area, Kafzeh 8 a contracted burial on its right side, Kafzeh 9 semi-flexed like number 8 and Kafzeh 10 an infant in a strongly flexed position. Finally some infant remains, Kafzeh 11, were recovered from Layer XXII from a grave containing some offerings.

Associated finds   The stone tools recovered from the Jebel Kafzeh site were of Levalloiso-Mousterian type; a rich industry containing denticulate tools, scrapers, Levallois points, backed knives and burins of Upper Palaeolithic character (Ronen and Vandermeersch, 1972).

The faunal remains found at this site include the remains of horse (*Equus*), rhinoceros (*Rhinoceros*), fallow deer (*Dama*), wild ox (*Bos*) and gazelle (*Gazella*), as well as some bird remains (Vandermeersch, 1966 and 1970). A more extensive faunal list resulted from the early excavation (Picard, 1937) and from a later study (Bouchud, 1971).

Dating   On the basis of the stratigraphy and the fauna the date of the Jebel Kafzeh has been given as the Last Pluvial of the Würm Glaciation (Vandermeersch, 1966) and as the end of the Last Interpluvial of the Würm Glaciation (Howell, 1959). No evidence has been offered so far as to the absolute date of the site, but some analytical evidence has been offered in respect of the relative dating (Oakley, Campbell and Molleson, 1975).

Morphology   The remains recovered from Jebel Kafzeh have been identified as follows:

Designation	Status	Bones	First reference
Kafzeh 1	Adult	Frontal	Neuville, 1934
Kafzeh 2	Adult	Calotte, mandible, teeth	Neuville, 1934
Kafzeh 3	Adult	Incomplete skeleton	Köppel, 1935
Kafzeh 4	Infant	Palate, mandible, teeth	Köppel, 1935
Kafzeh 5	Adult	Calotte, palate, teeth	Köppel, 1935
Kafzeh 6	Adult	Cranium, teeth	Köppel, 1935
Kafzeh 7	Adult	Cranial fragments, palate, teeth, mandible, clavicle, phalanges	Köppel, 1935
Kafzeh 8	Adult	Incomplete skeleton	Vandermeersch, 1966
Kafzeh 9	Adult	Incomplete skeleton	Vandermeersch, 1969
Kafzeh 10	Infant	Skeleton	Vandermeersch, 1969
Kafzeh 11	Child (10)	Skeleton	Vandermeersch, 1970

SKULL

The best preserved skull, Kafzeh 6, has been carefully restored and described (Vallois and Vandermeersch, 1972). It appears to be that of a young adult male on the basis of its skull form and the state of its teeth and sutures. The skull is a little distorted and some surface details are missing as well as much of the skull base, but sufficient remains of both vault and face to permit a satisfactory reconstruction.

The skull is generally robust and large with a high well-rounded vault, a modest supraorbital torus and no supratoral groove. The occipital bone is full and rounded with a low position of the inion and no evidence of an occipital chignon. In general form the neurocranium is long and broad with little or no postorbital waisting, well defined parietal bosses high on the bones and well marked mastoid processes that are broad based. The face is large and low with quadrangular orbits that slope downwards and laterally; the nasal cavity appears large both below and above giving rise to marked interorbital separation. The maxillae are large with canine fossae and well defined maxillozygomatic angulation. In

98

lateral view the face is virtually orthognathic with little or no projection of the central region of the face. In basal view the skull shows a wide shallow digastric groove, an elongated narrow and deep temporomandibular glenoid and a prolonged post-glenoid process. The palate is wide and shallow. Two middle ear ossicles from the site have also been described (Arensberg and Nathan, 1972).

MANDIBLE

The mandibular remains (Kafzeh 8 and 9) suggest that the jaws were tall and robust without trace of a *planum alveolare* but with well marked chins. The digastric fossae are placed on the internal aspect of the symphysial region and face both downwards and backwards.

TEETH

The Kafzeh 6 palate lacks both right premolars, the right lateral incisor and both third molars. In general all the teeth are large and heavily worn with dentinal exposure in all but the second molars. The plane of incisal wear is oblique and the incisors are not shovelled. The premolars are bicuspid, the first molars have four cusps including a hypocone but the second molars lack a hypocone and are consequently tritubercular. The right third molar possessed three roots. No details of the lower teeth are available.

POSTCRANIAL BONES

Up to the present no details of the postcranial bones are available.

Dimensions *Vallois and Vandermeersch* (1972)
Kafzeh 6

Max. Length 196   Max. Breadth 145·5
Cranial Index 73·7   Cranial Capacity 1568 cc

TEETH

Kafzeh 6		I^1	I^2	Upper Teeth (Crown Dimensions) C	PM^1	PM^2	M^1	M^2	M^3
Right	l	10·0	—	8·5	—	—	11·0	10·5	—
side	b	—	—	10·0	—	—	12·5	13·0	—
Left	l	10·5	8·5	8·5	7·0	6·5	11·0	10·5	—
side	b	—	8·0	10·0	10·0	10·0	13·0	13·0	

The Jebel Kafzeh Remains

Affinities   While it is true that the skeletons from Jebel Kafzeh are largely undescribed, sufficient information would seem to be available to permit some assessment of the affinities of the group as a whole. The description of the Kafzeh 6 skull (Vallois and Vandermeersch, 1972) underlines the preliminary opinion given in the earlier publications that, although the material shows some archaic features, its affinities lie nearer to modern *Homo sapiens* than to the classic Neanderthal variety of this species. Other authors who concur with this view include Howells (1974), and Stringer (1974) but Thoma (1965) regards the Kafzeh remains as Proto-Cro-Magnon in the same way that he regards the Skhūl population. Thoma sees in the latter remains signs of hybridization that occur again in the Jebel Kafzeh remains but this time the Cro-Magnon character of the skulls is, to him, even more pronounced. Only Brose and Wolpoff (1971) would include Jebel Kafzeh in a 'non-classic' Neanderthal group that comprises a wide range of African, Asian and Near Eastern fossil men. This viewpoint has been severely criticized (Howells, 1974).

The significance of the Jebel Kafzeh site rests principally, therefore, on the combination of the supposed date (Last Glaciation), the tool culture (Levalloiso-Mousterian) and the taxonomy of the remains (*Homo sapiens sapiens*), contradicting an older view that only Neanderthal man was responsible for Mousterian tools.

When further work reveals the details of the remaining material it will be possible to assess more closely the relationships of the Jebel Kafzeh remains to those from the other Near Eastern sites such as Skhūl, Tabūn, Amud (*q.v.*) and Shanidar. In turn this should throw some light on the Neanderthal problem as a whole.

Originals   Laboratoire de Paléontologie des Vertébrés et de Paléontologie Humaine. 4 Place Jussien, 75230, Paris.

Casts   Not available at present.

The Jebel Kafzeh Remains

References

NEUVILLE, R. 1934
Le préhistorique de Palestine. *Rev. Bibliq.* *43*, 249.

KÖPPEL, R. 1935
Das Alter der neuentdeckten Schädel von Nazareth. *Biblica 16*, 58–73.

NEUVILLE, R., 1936
Excavations in Palestine 1934-5. *Q. Dep. Antiqs. Palest. 5*, 199.

PICARD, L., 1937
Inferences on the problem of the Pleistocene climate of Palestine and Syria drawn from flora, fauna and stratigraphy. *Proc. Prehist. Soc. 3*, 58–70.

NEUVILLE, R. 1951
Le Paléolithique et le Mésolithique du désert de Judée. *Archs Inst. Paléont. Hum. 24*, 179–184.

HOWELL, F. C. 1959
Upper Pleistocene stratigraphy and early man in the Levant. *Proc. Am. Phil. Soc. 103*, 12–13.

THOMA, A. 1965
La définition des Néanderthaliens et la position des hommes fossiles de Palestine. *Anthropologie, 69*, 519–534.

VANDERMEERSCH, B. 1966
Nouvelles découvertes de restes humains dans les couches Levalloiso-Moustériennes du gisement de Qafzeh (Israël). *C.R. Acad. Sci. Paris. 262*, 1434–1436.

VANDERMEERSCH, B. 1969
Les nouveaux squelettes moustériens découverts à Qafzeh (Israël) et leur signification. *C. R. Acad. Sci. Paris., 268*, 2562–2565.

VANDERMEERSCH, B. 1970
Une sépulture moustérienne avec offrandes découverte dans la grotte de Qafzeh. *C. R. Acad. Sci. Paris. 270*, 298–301.

VANDERMEERSCH, B. 1971
Récentes découvertes de squelettes humains à Qafzeh (Israël): essai d'interprétation. In *Origine de l'homme moderne (Écologie et conservation 3)*, 49–54: UNESCO.

BOUCHUD, J. 1971
Étude préliminaire de la faune du Djebel Qafzeh prés de Nazareth (Israël). In *Proc. VIIIᵉ Congrès INQUA*. Paris: Assoc. Franç. pour l'Étude du Quaternaire.

BROSE, D. S. and WOLPOFF, M. H. 1971
Early Upper Paleolithic man and Late Middle Pleistocene tools. *Am. Anthrop. 73*, 1156–1194.

ARENSBERG, B. and NATHAN, H. 1972
A propos de deux osselets de l'oreille moyenne d'un néanderthaloide trouvés à Qafzeh. *Anthropologie 70*, 301–307.

RONEN, A. and VANDERMEERSCH, B. 1972
The Upper Paleolithic in the cave of Qafzeh (Israel). *Quaternaria 16*, 189–202.

VALLOIS, H. V. and VANDERMEERSCH, B. 1972
Le crâne mousterien de Qafzeh (Homo VI), *Anthropologie 76*, 71–96.

HOWELLS, W. W. 1974
Neanderthals: names, hypotheses and scientific method. *Am. Anthrop.*
76, 24-38.
STRINGER, C. B. 1974
Population relationships of Later Pleistocene hominids: a multi-variate study of available crania. *J. Archaeol. Sci. 1*, 317-342.
OAKLEY, K. P., CAMPBELL, B. G. and MOLLESON, T. I. 1975
Catalogue of fossil hominids Part III: Americas, Asia, Australasia.
London: Trustees of the British Museum (Natural History).

The Amud Remains

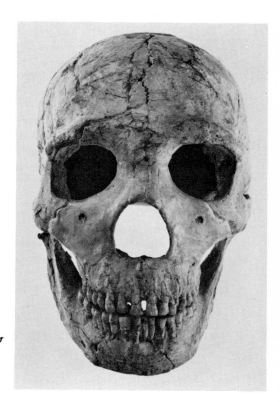

Fig. 30   The Amud I Skull
Frontal view
*Courtesy of Professor
Hisashi Suzuki*

*Synonyms
and other names*
Amud I–IV; Amud man

Site
A cave site on the lower course of the Wadi Amud, 10 km. north of the town of Tiberias, 50 km. east–north–east of Haifa. The cave is situated 3 km. from the mouth of the wadi which opens into Lake Tiberias.

Found by
Tokyo University Scientific Expedition to Western Asia (28th June–17th July, 1961). Director: H. Suzuki.

Geology
The deposits in the Amud area consist of Upper Cretaceous and Eocene limestones the upper part of which is exposed in the lower part of the Amud gorge. Around the cave the

103

limestone is overlain by veneers of gravels, basaltic lava
flows, lacustrine silts, sands and other non-marine sediments.
These are probably of Quaternary or Neogene age. The
slopes nearby are covered with a calcareous layer termed the
Nari crust. The stratigraphy has been disturbed in places by
the tectonic activity of the Rift valley. The limestone series is
divided into four from above downwards; 'Massive' limestone, 'Bedded' limestone, 'Weakly Bedded' limestone and
'Irregularly Bedded' limestone. Below these layers there is
chalk with layers of flint. The Amud cave opens between the
'Massive' and 'Bedded' layers and comprises a semicircular
depression about 12 m. by 10 m. the outer half of which is
covered by cave deposits while the inner half is exposed
limestone. The cave deposits consist of two layers separated by
an erosional unconformity. Layer A is recent and contains
potsherds but layer B is made up of loose calcareous silts with
limestone rubble. Layer B comprises strata (1–4) from above
downwards and Amud I was found as a contracted burial
just below the top of layer B1.

Associated finds Layer B is said to be a palaeolithic horizon representing a single
industrial cycle throughout which there is a mixture of
Levalloiso-Mousterian and Upper Palaeolithic stone tools.
The former group includes retouched points and racloirs
while the latter includes end-scrapers and burins. The form
and proportions of these tools may indicate a transitional or
intermediate industry between the Middle and Upper
Palaeolithic industries of the region.

The mammalian faunal remains are scanty and fragmented but
fossils representing gazelle (*Gazella cf. subgutturosa*), fallow
deer (*Dama mesopotamica*), wild ox (*Bos sp.*), wild pig (*Sus sp.*),
wild goat (*Capra sp.*) and horse (*Equus sp.*) have been found
as well as some birds and reptiles (Suzuki and Takai, 1970).

Dating Consideration of the fauna and the tools recovered from the
Amud site indicate a date within the Interstadial between the
Early and Main Fourth Glaciation (Würm).

Physical methods that support this date include uranium/
ionium growth which gives a date of 27,000±500 years
B.P. and uranium fission track which gives a date of 28,000   104

The Amud Remains

±35% years B.P. Radiocarbon dates range from 5,710±80 years B.P. for Upper B1 to 18,300±400 years B.P. for Basal B4. These dates are believed to be too 'young' as the result of contamination by younger carbon (Suzuki and Takai, 1970).

Morphology Amud I is the skeleton of an adult male of approximately 25 years of age. The skull was badly crushed laterally and the postcranial bones were highly fragmented. Amud II consists of an adult right maxillary fragment and Amud III and IV are skull fragments of infants.

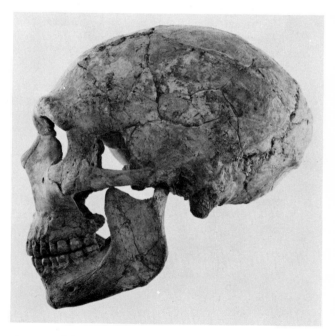

Fig. 31 The Amud I Skull
Left lateral view
Courtesy of Professor Hisashi Suzuki

AMUD I SKULL
This skull has been reconstructed using the Shanidar material to assist with missing areas. The vault is almost complete but much of the base is missing; the dentition is complete but

a good deal of the palate and the central part of the face are missing. The mandible is intact.

The vault of the skull is rounded with a receding forehead, a divided supraorbital torus, prominent mastoid processes and a rounded occipital bone. The cranial capacity, by water displacement of the cavity endocast, is very large (1,740 cc.) indeed it is the largest known capacity of all the fossil hominids.

The face is long and narrow, the vault is long and wide and the mandible is long as well as having a very large bicondylar breadth. The dental arcade is U-shaped and rather short and the mandibular symphysial region does not recede but neither is there a well marked chin.

TEETH

Both the upper and lower dentitions are complete (thirty two teeth in all) but there has been a good deal of post-mortem distortion of the tooth alignment. In lateral view the occlusal plane appears to be almost flat.

The incisors and canines are broad labio-lingually with well developed lingual tubercles. The premolars are bicuspid and the second premolars (PM2) are larger than the first premolars (PM1) in both upper and lower jaws. The first and second upper molars have large hypocones and the mesiodistal lengths of the lingual sides are longer than those of the buccal sides. Both upper molars are reduced in size.

The lower first molars have a Y5 cusp pattern while the lower second molar on the left has a +4 pattern and the second molar on the right an X4 pattern. Both lower third molars have five cusps and a median lingual accessory cuspule.

An isolated molar was also recovered from the site.

POSTCRANIAL BONES

Loss of the ends of most of the limb bones renders length assessments and therefore height estimates hazardous. However, a standing height of 172·3–177·8 cm. has been given. In general the long bones are long and slender, the hand is large with large joints and the pelvis (although badly broken) appears narrow.

The Amud Remains

Suzuki and Takai (1970)

AMUD I SKULL

Max. Length 215   Max. Breadth 154
Cranial Index 71·6   Cranial Capacity 1,740 cc
(Dolichocephalic)

AMUD I MANDIBLE

Length 119   Bicondylar Breadth 145

AMUD I TEETH

| | Upper Teeth (Crown Dimensions) | | | | | | | |
	I^1	I^2	C	PM^1	PM^2	M^1	M^2	M^3
Right l	9·3	7·7	8·5	7·3	6·6	10·7	10·4	6·8
Side b	8·2	8·4	9·5	10·4	10·1	12·4	12·3	7·7
Left l	—	7·7	8·5	7·4	6·6	10·7	10·4	8·5
Side b	—	8·4	9·5	10·6	10·0	12·5	12·2	11·0

| | Lower Teeth (Crown Dimensions) | | | | | | | |
	I_1	I_2	C	PM_1	PM_2	M_1	M_2	M_3
Right l	—	6·3	—	(7·5)	6·9	(11·0)	11·3	11·6
Side b	—	7·5	—	9·1	8·5	10·8	10·3	10·5
Left l	(5·0)	(6·1)	7·6	(7·4)	(6·5)	(10·9)	(10·5)	11·8
Side b	(7·2)	7·5	9·1	8·9	8·5	10·8	10·8	10·8

AMUD I POSTCRANIAL BONES

Femur Maximum Estimated Length 489
Tibia Maximum Estimated Length 386
( ) Approximate measurement

Affinities   The overall characteristics of the Amud remains seem to suggest that they are part of the Near Eastern Neanderthal group showing a mixture of West Asian features and those of Upper Palaeolithic man. Amud man is said to be closest to the Shanidar and Tabūn remains (*q.v.*) but a little more advanced, yet still showing affinities to the Skhūl and Kafzeh remains (*q.v.*) (Suzuki and Takai, 1970). Howells (1974) includes Amud as a Near Eastern Neanderthaler with Tabūn and

The Amud Remains

Shanidar while Stringer (1974) emphasizes a closer relationship of the Amud remains to Upper Palaeolithic material than to the 'classic' Neanderthalers.

Until more opinions or more material are available it seems that Amud man was part of the evolving Near Eastern populations displaying a mixture of morphological characters some of which are best compared with those of 'classic' Neanderthalers while others are reflected in later Upper Palaeolithic material. The smallness of the total available sample would seem to preclude a more precise evaluation at the present time.

Originals Amud I is held at the Rockefeller Museum, Jerusalem.

Casts Not generally available at present.

References   VALLOIS, H. V. 1962
Un nouveau Néanderthaloide en Palestine. *Anthropologie 66*, 405-406.
SUZUKI, H. and TAKAI, F. (Eds.) 1970 *The Amud Man and his cave site.*
Tokyo: Keigaku Publishing Co.
HOWELLS, W. W. 1974
Neanderthals: names, hypotheses and scientific method. *Am. Anthrop.*
76, 24-38.
STRINGER, C. B. 1974
Population relationships of Later Pleistocene hominids: a multi-variate study of available crania. *J. Archaeol. Sci. 1*, 317-342.

The Neanderthal Problem

From the time of the first recognition of Neanderthal man his precise evolutionary position has been a source of debate and the discussion still continues without clear resolution. One of the longstanding puzzles of the Neanderthal story has been their supposed sudden disappearance from the fossil record, a suggestion that was perhaps more plausible in the earlier days when only 'classic' Neanderthalers were known from European sites and when the Mousterian culture was firmly associated with the Neanderthalers alone. The 'catastrophic' theory of Neanderthal disappearance has been supported by suggestions of epidemics, of conflicts with more advanced peoples, of changes in the climate that left the cold-adapted Neanderthaler vulnerable to new diseases and of course, of absorption into the gene pool of incoming migratory peoples. It seems that the Aurignacian culture arrived in Europe fully formed and that Aurignacians and Mousterians were contemporaneous in parts of Europe from about 40,000 years B.P. However, the debate between archaeologists on tool variability should inspire caution regarding the correlation of forms of man and specific cultures. If the Neanderthalers did disappear suddenly, and the idea has been forcefully attacked (Brace, 1964), there seems no imperative reason to look for a single cause; a combination of circumstances is possible, even reasonable.

Currently the phylogeny of Neanderthal man can be summarized in four differing ways, each with its supporters.

I. THE PRESAPIENS HYPOTHESIS
This view holds that the Neanderthalers form a side branch of the main human evolutionary line and became extinct at the end of the Early Würm glaciation. At the same time the fossils of the main line proceeded towards modern man and form the Presapiens group (Vallois, 1954; Weiner, 1958; Thoma, 1965; Leakey, 1972).

2. THE PRE-NEANDERTHAL HYPOTHESIS
In the early Upper Pleistocene 'generalized' or 'progressive' Neanderthals of Europe and the Middle East are recognized and it is suggested that in the Last Glaciation (Würm) some populations became cut off in the West European peninsula.

These Neanderthals are regarded as becoming specialized for cold, and severe natural selection plus a restricted gene flow led to 'classic Neanderthal' isolates exemplified by La Chapelle, La Ferrassie, Neanderthal and many others. At the same time 'generalized' groups such as Tabūn, Skhūl, Steinheim and Swanscombe, developed the characteristics of modern man (Sergi, 1953; Howell, 1957; Breitinger, 1957; and Le Gros Clark, 1966).

3. THE NEANDERTHAL PHASE OF MAN HYPOTHESIS
This view supports a single lineage arising from *Homo erectus*, by successive evolution, through a Neanderthal phase to modern man (Schwalbe, 1904; Hrdlička, 1930; Weidenreich, 1943 and 1949; and more recently, Brace, 1964; Brose and Wolpoff, 1971; Mann and Trinkaus, 1973).

4. THE SPECTRUM HYPOTHESIS
A view that arose out of a reappraisal of the Swanscombe skull by Weiner and Campbell (1964) who emphasized the interrelatedness of early forms of man and suggested a 'Neanderthaloid Intermediate' group including Steinheim, Swanscombe, Skhūl V and Krapina. All these as well as some others were regarded as a 'spectrum of varieties'.

Clearly the question of the definition of terms is crucial here, and the literature reveals great confusion. Hrdlička (1930) took a cultural view stating that Neanderthal man was 'the man and period of the Mousterian culture'. Brace (1964) added to this later by stating that Neanderthalers were 'the men of the Mousterian culture prior to the reduction in size and form of the Middle Pleistocene face'. Both of these definitions are open to criticism since associating cultural traditions with specific hominid categories is hazardous—some tool sites have no hominids and some hominid sites no tools. In any event the Mousterian culture is complex and typologically classified by some into several sub-divisions yet some archaeologists prefer to regard tools as a series of functional assemblages.
Morphological definitions given in the past by Le Gros Clark (1966), Thoma (1965), Boule and Vallois (1957) and Vandermeersch (1972) would limit the use of the term

Neanderthal to Western European examples of 'classic' morphology. Brose and Wolpoff (1971), however, give a temporal and morphological definition that includes 'all hominid specimens' within the time span of the end of the Riss to the appearance of anatomically modern *Homo sapiens*. That is a 'classic cold' group from Western Europe and all others regarded as Neanderthaloid. This view has been severely criticized by Howells (1974) on the grounds that they have failed to comprehend the need to seek distinctions in levels of within-group and between-group variations in statistical terms. Howells is quite clear in his view of only using the term 'Neanderthal' to include the European Würm skulls ('classic') plus Tabūn, Shanidar and Amud. He excludes Skhūl, Kafzeh, Ighoud and Petralona as well as the Rhodesian and Solo finds.

In general Stringer (1974) is in agreement with Howell's view since his multivariate study was on crania alone and made use of Howell's comparative data. Stringer favours the morphological definitions of Morant (1927), Hrdlička (1930) and Le Gros Clark (1966), since his cranial study supports their anatomical opinions. In addition the same study has cast doubts on the existence of a close relationship between the 'classic' Neanderthaler and the Upper Palaeolithic populations while showing that early Neanderthal crania were morphologically more like modern *Homo sapiens* than were the later examples.

To summarize it seems clear that the *Pre-sapiens Hypothesis* of the very early appearance of modern man, as exemplified by Swanscombe and Fontéchevade is not widely supported at present. However, the derivation of Upper Palaeolithic and modern man through a unilinear scheme of *in situ* succession including the later Neanderthals is not without its critics. The complexity of relationships shown by the multivariate statistical studies of Howells and Stringer should caution those who would argue for a *Neanderthal Phase of Man*. Thirdly, placing the material into *Homo sapiens* yet creating geographic subspecific categories for each find as in the *Spectrum Hypothesis* may simplify nomenclature but it does little to clarify phylogeny. Finally, the *Pre-Neanderthal Hypothesis*

The Neanderthal Problem

will need further support before it can be fully accepted. This support could come in the form of evidence of isolation and restricted gene flow and further evidence of cold adaptation in the 'classic' types and more examples of the lineage of early generalized forms that may have given rise to modern man by escaping the trap of Neanderthal specialization.

References

SCHWALBE, G. 1904
Die Vorgeschichte des Menschen. Braunschweig: Friedrich Vieweg und Sohn.

MORANT, G. M. 1927
Studies of palaeolithic man II. A biometric study of Neanderthaloid skulls and of their relationships to modern racial types. *Annls. Eugen. Lond. 2*, 318-380.

HRDLIČKA, A. 1930
The skeletal remains of early man. *Smithson. Misc. Coll. 83*, 1, 1-379.

WEIDENREICH, F. 1943
The 'Neanderthal Man' and the ancestors of 'Homo sapiens'. *Am. Anthrop. 42*, 375-383.

WEIDENREICH, F. 1949
Interpretations of the fossil material. In *Ideas on human evolution.* Ed. W. W. Howells, 1962. Cambridge, Mass.: Harvard University Press.

SERGI, S. 1953
Morphological position of the 'Prophaneranthropi' (Swanscombe and Fontéchevade). *R. C. Accad. Lincei 14*, 601-608.

VALLOIS, H. V. 1954
Neandertals and praesapiens. *J. R. Anthrop. Inst. 84*, 111-130.

BOULE, M. and VALLOIS, H. V. 1957
Fossil Men. London: Thames and Hudson.

BREITINGER, E. 1957
On the phyletic evolution of Homo sapiens. In W. W. Howells, Ed. 1962. *See* WEIDENREICH 1949 *supra.*

HOWELL, F. C. 1957
The evolutionary significance of variations and varieties of 'Neanderthal' man. *Quart. Rev. Biol. 32*, 330-347.

WEINER, J. S. 1958
The pattern of evolutionary development of the genus Homo. *S. Afr. J. Med. Sci. 23*, 111-120.

BRACE, C. L. 1964
The fate of the 'Classic' Neanderthals: a consideration of hominid catastrophism. *Curr. Anthrop. 5*, 3-43.

WEINER, J. S. and CAMPBELL, B. G. 1964
The taxonomic status of the Swanscombe skull. In C. D. Ovey, Ed. *The Swanscombe Skull. A survey of research on a Pleistocene site.* London: Royal Anthrop. Inst.

THOMA, A. 1965
La définition des Néandertaliens et la position des hommes fossiles de Palestine. *Anthropologie 69*, 5-6; 519-534

CLARKE, W. E. LE GROS 1966
The fossil evidence for human evolution. Chicago: Chicago University Press.

BROSE, D. S. and WOLPOFF, M. H. 1971
Early upper palaeolithic man and late middle palaeolithic tools. *Am. Anthrop. 73*, 1156-1194.

LEAKEY, L. S. B. 1972
Homo sapiens in the Middle Pleistocene and the evidence of Homo sapiens' evolution. In F. Bordes, Ed. The Origin of Homo Sapiens. *Proceedings of the Paris Symposium, U.N.E.S.C.O. (INQUA)*, 25-29.

VANDERMEERSCH, B. 1972
Recentes découvertes de squelettes humains à Qafzeh (Israël): essai d'interpretation. In. F. Bordes, Ed. 1972 *supra*.

MANN, A. and TRINKAUS, E.
Neandertal and Neandertal-like fossils from the Upper Pleistocene. *Yearb. Phys. Anthrop. 17*, 169-193.

HOWELLS, W. W. 1974
Neanderthals: names, hypotheses and scientific method. *Am. Anthrop. 76*, 24-38.

STRINGER, C. B. 1974
Population relationships of later pleistocene hominids: a multivariate study of available crania. *J. Archaeol. Sci. 1*, 317-342.

North-west Africa

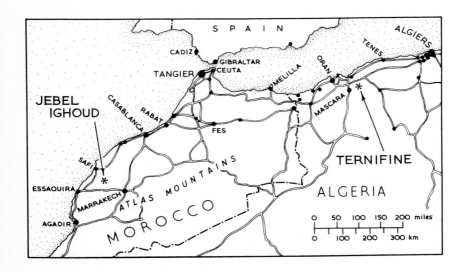

Fig. 32   Hominid fossil sites in North-west Africa

The Ternifine Remains

Synonyms and other names	*Atlanthropus mauritanicus* (Arambourg, 1954) Ternifine man
Site	Ternifine, near the village of Palikao, 11 miles east of Mascara, Oran, Algeria.
Found by	C. Arambourg and R. Hoffstetter, July, 1954; C. Arambourg, 1955.
Geology	Commercial excavation of a large hill of sand exposed the bed of a Pleistocene artesian lake containing fossil bones and stone tools. The level of the water table made it necessary for pumps to be used in the recent exposure of the lower layers.
Associated finds	Several hundred implements were recovered from the site, made principally of quartzite and sandstone but occasionally of poor-quality flint. The tools included primitive biface hand-axes, scrapers and a type of small axe with a curved blade; an industry described as Acheuléen II (Balout and Tixier, 1957). The associated fossil fauna was rich and varied containing many extinct forms suggestive of a tropical savannah environment. The predominant mammals were hippopotamus (*Hippopotamus amphibius*), elephant (*Loxodonta atlantica*), early zebra (*Equus mauritanicus*), rhinoceros (*Ceratotherium simum-mauritanicum*), camel (*Camelus thomasi*), several antelopes and numerous carnivores. In addition, several species of particular importance were identified, a sabre-toothed cat (*Homotherium latidens*), giant wart-hog (*Afrochoerus sp.*) and a giant baboon (*cf. Simopithecus*).
Dating	The character of the fauna, especially the presence of the last three species, in conjunction with the type of industry, allowed Arambourg (1955a) to establish an early Middle Pleistocene date for these deposits.
Morphology	*Arambourg* (1963) The hominid bones which were discovered comprise three mandibles and a parietal bone.

MANDIBLES

The mandibles are all adult and remarkable for their size and robustness. Nos. I and III are probably male, whereas No. II despite its great size is probably female. In all three specimens **117**

Fig. 33   The Ternifine I mandible
Right lateral view
Courtesy of the late Professor C. Arambourg

Fig. 34   The Ternifine I mandible
Occlusal view
Courtesy of the late Professor C. Arambourg

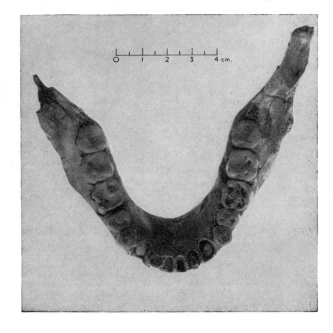

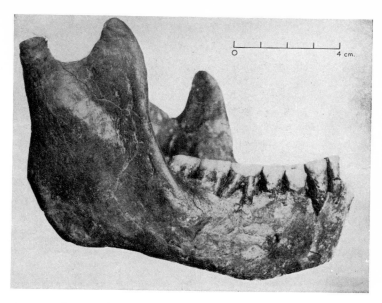

Fig. 35 The Ternifine III mandible
Right lateral view
Courtesy of the late Professor C. Arambourg

Fig. 36 The Ternifine III mandible
Occlusal view
Courtesy of the late Professor C. Arambourg

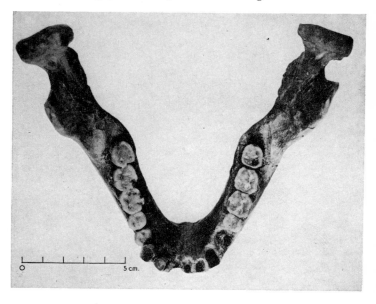

the borders of the bodies of the mandibles are parallel, the symphyses slope backwards, so that there are no chins, but a sub-mental trigone is present in Nos. II and III. The dental arcades are parabolic and there are several mental foramina in Nos. I and III. The body of No. I is strengthened by a prominent marginal torus. The rami are broken, but enough remains to suggest some variability; none the less they are all broad with marked impresses for masseter and medial pterygoid muscles.

TEETH

The incisors and canines are known only from the two worn specimens present in No. III. Little more can be said than that they appear hominid. The premolars are large, the first having an asymmetrical crown with a prominent buccal cusp and a smaller lingual cusp more distally placed. The second premolars have large buccal and lingual cusps placed opposite to each other, and large posterior fovea. The premolar teeth have large labial cingula whilst the second premolar, which is little worn, shows marked secondary enamel wrinkling.

The molar teeth are large and decrease in size in the order $M2 > M1 > M3$. The cusp pattern is dryopithecine (Y5 or + 5),

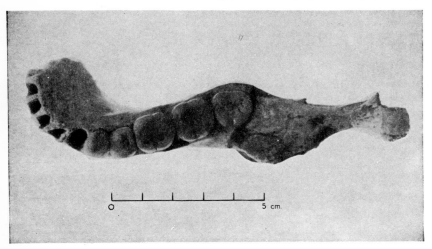

Fig. 37   The Ternifine II mandible
Occlusal view
Courtesy of the late Professor C. Arambourg

120

The Ternifine Remains

sixth cusps being present in No. II (M2, M3) and No. III (? M3) (Howell, 1960). It is probable that secondary enamel wrinkling was present on the occlusal surfaces of the molars as shown by those which are least worn; signs of a buccal cingulum are present on some of the molar teeth.

THE PARIETAL (Arambourg, 1955c)

The parietal is young, for all its sutures were open at the time of death. Its thickness is within the range of variation found in modern man. The parietal curve indicates that the vault was probably low and that the widest diameter of the cranium was bitemporal rather than biparietal. The temporal lines are well inscribed and testify to the power of the temporalis muscle. Internally the bone is heaped to form a prominent Sylvian crest whilst the middle meningeal pattern is primitive in its design, the principal division of the vessel having occurred before it reached the parietal bone. Thereafter, the anterior and bregmatic branches are weak but the temporal branch is stronger.

Dimensions   *Arambourg (1955b) and Howell (1960)*

Mandibles	No. I	No. II	No. III
Symphysial height	39	35	39
Body height (Behind M1)	35	34	38
Body thickness (Behind M1)	19	16	20
Ramus height	—	72	93
Ramus breadth	—	45	48
Mandibular angle	—	98°	111°
Total length	110	110	129
Bicondylar breadth	—	—	158

Mandibles	Lower Teeth (Crown Dimensions)		
	No. I	No. II	No. III
PM1 l	8·5	9·0	8·0
b	9·0	11·2	10·0
PM2 l	8·0	9·5	8·2
b	10·0	10·5	10·0
M1 l	12·8	14·0	12·0
b	12·0	13·0	11·8
M2 l	13·0	14·2	12·0
b	13·7	13·7	12·1
M3 l	12·0	13·2	8·0
b	12·5	12·5	11·5

The Ternifine Remains

Affinities   The similarities between the Ternifine mandibles and those of Choukoutien Locality I (*Homo erectus*) are particularly striking. Features such as the mandibular torus, the shape of the dental arcade and the mental trigone, the form of the first premolar, the molar size order, the enamel wrinkling and the presence of cingular ridges, all argue for their morphological relationship. These likenesses were recognized by Arambourg (1954), but the Ternifine remains have been provisionally named *Atlanthropus mauritanicus* until more is known of these African representatives of the pithecanthropines of the Far East. Howell (1960) also accepted the resemblances between the Ternifine and Choukoutien mandibles and asserted that 'there are no differences between the Ternifine and Choukoutien Locality I people which might not be expected within a single polytypic species, populations of which were widely separated geographically'.

Originals   Institut de Paléontologie, 8, Rue de Buffon, Paris–5ᵉ, France.

Casts   Not generally available at present.

References   ARAMBOURG, C., and HOFFSTETTER, R. 1954
Découverte, en Afrique du Nord, de restes humains du Paléolithic inférieur. *C. R. Acad. Sci. Paris 239*, 72-74.
ARAMBOURG, C. 1954
L'hominien fossile de Ternifine (Algérie). *C. R. Acad. Sci. Paris 239*, 893-895.
ARAMBOURG, C. 1955a
A recent discovery in human paleontology: *Atlanthropus* of Ternifine (Algeria). *Am. J. Phys. Anthrop. 13*, 191-202.
ARAMBOURG, C. 1955b
Une nouvelle mandibule 'd'Atlanthropus' au gisement de Ternifine. *C. R. Acad. Sci. Paris. 241*, 895-897.
ARAMBOURG, C. 1955c
Le pariétal de l'*Atlanthropus mauritanicus*. *C. R. Acad. Sci. Paris 241*, 980-982.
ARAMBOURG, C. 1956
Une IIIᵉᵐᵉ mandibule d' '*Atlanthropus*' découverte à Ternifine. *Quarternaria 3*, 1-4.
BALOUT, L., and TIXIER, J. 1957
L'Acheuléen de Ternifine. *C. R. Congr. préhist. Fr. XVᵉ, 1956*, 214-218.
HOWELL, F. C. 1960
European and northwest African Middle Pleistocene hominids. *Curr. Anthrop. 1*, 195-232.
ARAMBOURG, C. 1963
Le gisement de Ternifine. Part I C. Arambourg and R. Hoffstetter, Part II C. Arambourg. *Archs Inst. Paléont. hum. 32*, 1-190.

The Jebel Ighoud Remains

MOROCCO   JEBEL IGHOUD

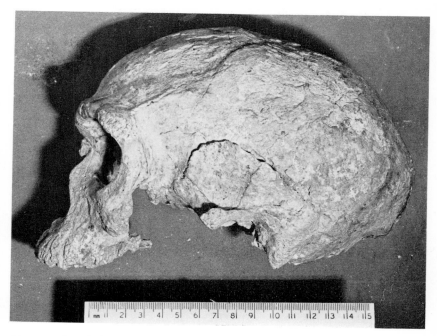

Fig. 38   The Jebel Ighoud I Skull
Left lateral view
Courtesy of Dr Christopher Stringer and the Musée de l'Homme, Paris

Synonyms   Ighoud 1–3; Ighoud man
and other names

Site   At a barytes mine, Jebel Ighoud, 60 km. south-east of Safi,
Morocco.

Found by   Ighoud 1. Workmen 1961, investigated by E. Ennouchi 1962.
Ighoud 2. Workmen and E. Ennouchi, 1968.
Ighoud 3. E. Ennouchi 1968.

Geology   The deposits containing the Jebel Ighoud remains were
fissure fillings within the Pre-Cambrian limestone of the   123

barytes mine; the filling material was made up of red clay, rock fragments and fossil bones. The fissure (8 m. in height by 5 m. in breadth) that contained the remains must have been originally a cave or a cavity open above by means of a narrow neck. The working gallery of the mine exposed the fissure and its filling in cross-section and the remains were found near the floor.

Associated finds Following the recovery of the second skull (Ighoud 2) an abundant flint industry was recovered whose character has been given as Levalloiso-Mousterian. Some rare quartzite implements have also been recovered as well as a number of eggshell fragments from the egg of a large bird. It has been suggested that the egg could have served as a receptacle.

The fauna that was recovered includes gazelles (*Gazella atlantica, G. cuvieri, G. dorcas*), rhinoceros (*Rhinoceros sp.*), horse (*Equus mauritanicus*), wild dog (*Canis anthus*), bovids (*Alcelaphus bubalis, Connochaetes taurinus, Bos ibericus, Bos primigenius*) and wild ass (*Asinus africanus*) (Ennouchi, 1962b).

Dating The only dating evidence that is available is that based on the faunal list and the stone tool descriptions given above. The conclusion drawn from this was that the site is 'indisputably' Middle Pleistocene (Ennouchi, 1963). However, Oakley (1964) has suggested a considerably younger age on the basis of sea level changes. No chemical, radiometric or other evidence has been put forward so far as to the date of the site other than an attempt to use the radiocarbon method. This resulted in the conclusion that the date of the site was beyond the capacity of this method (Ennouchi, 1968).

Morphology The remains identified from the Jebel Ighoud site comprise two skulls (Ighoud 1 and 2) and an immature mandible (Ighoud 3).

SKULL

The first skull lacks a mandible and all the upper teeth are broken off at the tooth necks. Much of the base is missing including the area of the foramen magnum. The skull is said to have belonged to an adult male of about 35 years of age and shows a continuous supraorbital torus surmounted by a supratoral sulcus. The frontal profile retreats and leads on to a generally flattened vault and a flattened lambdoid region. The face shows subnasal prognathism with a large palate and large temporomandibular glenoid cavities. The orbits are large, square and widely separated by a large nasal cavity. The second skull is a calvaria with all of the face below the brow ridge missing, as is much of the base. The left side retains much of the occipital bone and the left half of the foramen magnum is complete. In many respects the second specimen resembles the first but may have belonged to a younger individual (perhaps 20–25 years of age) and has a slight indentation of the supraorbital torus centrally and an 'open' metopic suture. The walls of both skulls are thick but the former specimen is more robust than the latter.

TEETH

There are no complete teeth present in either skull but from the retained roots and the spaces left for the crowns it appears that they may have been large. Both premolars of Ighoud 1 are double rooted (Ennouchi, 1962 a and b, 1968).

MANDIBLE

Ighoud 3 is a broken juvenile mandible containing both deciduous and permanent teeth. The body of the mandible is intact on the left but broken through at its junction with the ramus on the right. On both sides the first permanent molars are in place and on the left the second deciduous molar is also retained. The remaining deciduous teeth that had erupted are broken off and many of their roots are present in their alveoli. Skiagrams of the mandible disclose the germ of the left second permanent molar in its crypt and on the right the germs of the first and second permanent premolars (Ennouchi, 1969).

Dimensions Ennouchi (1962 a, b and c, 1969)

SKULLS

	Jebel Ighoud 1	Jebel Ighoud 2
Max. Length	198	197
Max. Breadth	145	148
Cranial Index	73·2	75·1
Cranial Capacity (Ighoud 1)	1,480 cc	(Displacement method)

TEETH (Ighoud 3)

	Lower Teeth (Crown Dimensions)							
	I_1	I_2	C	PM_1	PM_2	M_1	M_2	M_3
Left 1	—	—	—	—	—	14·6	—	—
side b	—	—	—	—	—	12·3	—	—
Right 1	—	—	—	—	—	14·5	—	—
side b	—	—	—	—	—	12·2	—	—

Left Second Deciduous Molar Length   11·7
Breadth 10·3

Affinities   The first opinion given by the finder of these remains was that they were Neanderthal in the classic sense and thus akin to La Chappelle and La Ferrassie (*q.v.*) (Ennouchi, 1962b), and in his subsequent publications there seems to be no basic change in that view. Piveteau (1967), however, does not fully accept that interpretation, neither does Howells (1973, 1974), both authors emphasizing 'modern' features in the skulls. Howells particularly notes the divergence in facial and occipital morphology between the Jebel Ighoud skulls and such Near Eastern Neanderthals as Tabūn, Shanidar and Amud (*q.v.*). Stringer (1974), after an extensive metrical analysis, concluded that the general relationships of the Ighoud material were closest to those of Saccopastore, Amud and Skhūl 5.

In summary the few opinions that have been given would seem to be against those of the finder in terms of relationship to the classic Neanderthals and favour slightly more 'modern' affinities.

Originals   Laboratory of Geology, University Mohammed V, Avenue Moulay-Cherriff, Rabat, Morocco.

Casts   Not generally available.

References   ENNOUCHI, E. 1962a
Un crâne d'Homme ancien au Jebel Irhoud (Maroc). *C. R. Acad. Sci. Paris* 254, 4330–4332.
ENNOUCHI, E. 1962b
Un Néanderthalien; L'Homme du Jebel Irhoud (Maroc). *Anthropologie* 66, 279–299.

ENNOUCHI, E. 1963
Les Néanderthaliens du Jebel Irhoud (Maroc). *C.R. Acad. Sci. Paris.* *256*, 2459-2460.

OAKLEY, K. P. 1964
Frameworks for dating fossil man. London: Weidenfeld & Nicolson.

PIVETEAU, J. 1967
Un parietal humain de la grotte du Lazaret (Alpes-Maritimes). *Annls Paléont. 53*, 165-169.

ENNOUCHI, E. 1968
Le deuxième crâne de L'Homme d'Irhoud. *Annls Paléont. (Vertébrés) 54*, 117-128.

ENNOUCHI, E. 1969
Présence d'un enfant Néanderthalien au Jebel Irhoud (Maroc). *Annls Paléont. (Vertébrés) 54*, 251-265.

HOWELLS, W. W. 1973
Evolution of the genus Homo. Reading, Mass.: Addison Wesley.

HOWELLS, W. W. 1974
Neanderthals: names, hypotheses and scientific method. *Am. Anthrop. 76*, 24-38.

STRINGER, C. B. 1974
Population relationships of Later Pleistocene hominids: a multivariate study of available crania. *J. Archaeol. Sci. 1*, 317-342.

East and Central Africa

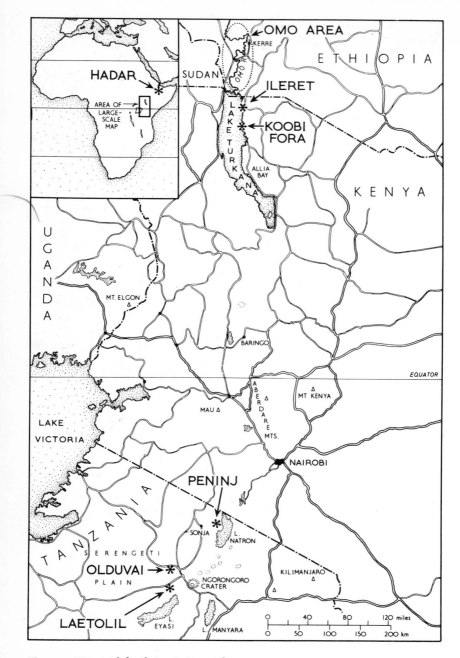

Fig. 39   Hominid fossil sites in East Africa

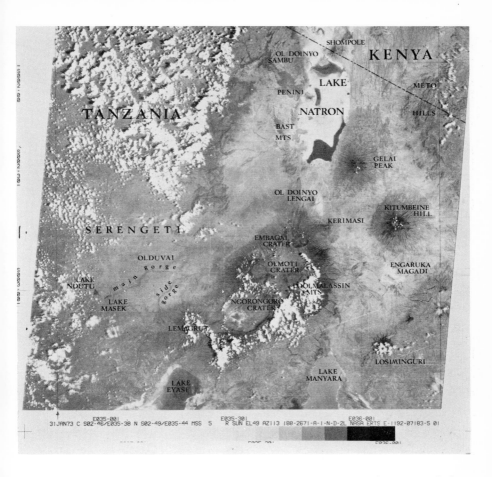

Fig. 40   Satellite view of Northern Tanzania showing geographic features, including
Olduvai Gorge
Courtesy of the National Aeronautical Space Administration, U.S.A. (Original image).
Reproduction by courtesy of the U.S. Geological Survey, Earth Resources Observation
Systems Data Center

131

The Olduvai Remains

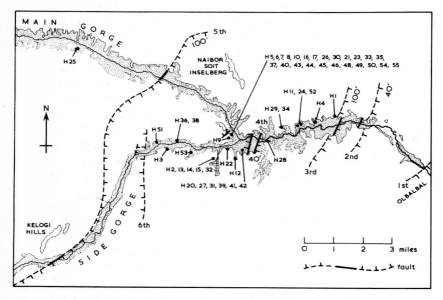

Fig. 41   A map of Olduvai Gorge indicating hominid find sites and the positions of geological faults in the area

Synonyms and other names

1. *Zinjanthropus boisei* (Leakey, 1959); *Paranthropus boisei* (Robinson, 1960); *Australopithecus* (*Zinjanthropus*) *boisei* (Leakey, Tobias and Napier, 1964); *Australopithecus boisei* (Tobias, 1967)

2. *Homo habilis* (Leakey, Tobias and Napier, 1964); *Homo africanus* (Robinson, 1972)

3. *Homo leakeyi* (Heberer, 1963); *Homo erectus* (Leakey, M.D., 1971a)

Site   Olduvai Gorge, Tanzania, East Africa, 110 miles south-west of Nairobi.

Geology   Olduvai Gorge★ is a canyon, in places as much as 300 ft. deep, about 30 miles long cut into the Serengeti Plain

★ Ol Duvai means 'The place of the wild sisal'; it still grows in the gorge.

132

The Olduvai Remains

north of Lake Eyasi and west of Ngorongoro Crater. Extending westwards from the Ol Balbal depression it forms a forked gorge cut solely by erosion. The fossiliferous strata, which rest on a lava base in the eastern part of the gorge, are bedded and consist of a series of lacustrine, fluviatile and wind-blown deposits some of which are volcanic tuffs. The strata have been numbered from below upwards I–IV followed by the Masek, Ndutu and Naisiusiu Beds. Thus Olduvai contains a stratigraphic record of about the past two million years in a rather small basin that has been the site of periodic deformation and faulting. A perennial saline lake was present during Bed I and Lower Bed II times, but it was reduced by earth movements after the deposition of Lower Bed II. Subsequently the lake was inconstant and moved progressively eastward.

Bed I is distinguished from Bed II by a flagstone layer chosen by Reck and Leakey in 1931 as a matter of convenience. For many years it appeared that this arbitrary line coincided with a major faunal change; it is now clear that geologically, faunally and culturally, Lower Bed II belongs with Bed I.

Since 1935 the site has been excavated by the late Dr L. S. B. Leakey and his wife, Dr M. D. Leakey; it has produced a prodigious quantity of mammalian fossils and stone tools. The most interesting bones, from an anthropological point of view, are those of fossil hominids. These hominids have been numbered and a complete list of them is included as a table with some details of the relevant stratum, locality, taxonomic attribution and associated industry. The map shows the position of some of the find-sites as well as the major faults. The geology of the Olduvai area has been described in detail (Hay, 1963 and 1976).

Found by Apart from Olduvai Hominid 1, which was found by H. Reck in 1913, all of the Olduvai Hominids have been found by the late Dr L. S. B. Leakey or by Dr M. D. Leakey and their co-workers since 1935.

Associated finds STONE TOOLS
Bed I (Leakey 1966, 1971a)
The stone tools recovered from this layer belong to an industry termed the Oldowan. It is best known for its choppers, primitive tools made by flaking one edge of a   133

water worn cobble to form a jagged cutting edge. There are some other small tools such as scrapers and chisels that were used for lighter work.

Bed II (Leakey 1971a)
Two differing industries occur in Bed II, the Developed Oldowan and the Acheulean. The Developed Oldowan is directly evolved from the Oldowan of Bed I and has the same tool-types with the addition of a few small and crudely made hand-axes.
In the Acheulean the characteristic tools are hand-axes and cleavers. They are made from large flakes and resemble the hand-axes found in Bed IV. The Developed Oldowan is generally found in the western part of the gorge while the Acheulean sites are on the alluvial fan to the east.

Bed III
The artefacts recovered from Bed III at site J.K.2 include hand-axes, heavy duty picks, choppers, small scrapers and borers as well as cores and flakes (Kleindienst, 1973).

Bed IV
Two contemporary industries have been recovered from Bed IV. One appears to be a derivative of the Developed Oldowan from Bed II and consists of relatively few small and rather poorly made hand-axes. The Acheulean variant consists of boldly flaked cleavers and hand-axes made from large flakes with little secondary flaking. These tools are very similar to the Lower Acheulean cleavers and hand-axes from Bed II (Leakey, 1971b).

Masek Beds
One site at Olduvai represents the latest phase of the Acheulean in this area. The characteristic tools are very large hand-axes usually made of quartzite very finely trimmed. In addition there are many small scrapers but no cleavers.

Ndutu Beds
The tools from this bed are Middle Stone Age tools made by a prepared core or Levallois technique recognized from many parts of Africa, Europe and elsewhere.

134

The Olduvai Remains

Naisiusiu Beds
The tools from these beds are relatively recent and consist of small crescents and other geometric forms made from chert or obsidian.

FAUNAL REMAINS
Bed I and Lower Bed II
The fossil mammalian fauna recovered from Bed I and Lower Bed II is extensive. It includes two proboscidians (*Deinotherium bozasi* and *Elephas recki*), several archaic pigs (including *Mesochoerus sp.*, *Notochoerus sp.*, *Potamochoerus sp.* and *Metridiochoerus sp.*), several bovids and giraffids, and many other forms such as rodents, carnivores, primates and equids. There were also numerous remains of small amphibia, reptiles and fish (Leakey, 1965).

Upper Bed II
The mammalian fauna of Upper Bed II contains many giant herbivores including hippopotamus (*Hippopotamus gorgops*), pigs (*Stylochoerus nicoli*, *Mesochoerus olduvaiensis*, *Potamochoerus majus*), a giant bovid (*Pelorovis oldowayensis*), equids (*Equus oldowayensis*, *Stylohipparion albertense*); also large carnivores, rhinoceroses and primates (Leakey, 1965).

Bed III
The fauna known from Bed III includes hippopotami, bovids and equids. In addition there were Lagomorpha, Rodentia, Carnivora, Perissodactyla and Proboscidea (Kleindienst, 1973).

Bed IV
Most of the animals found in Bed III and upper Bed II persist into Bed IV, but crocodiles and catfish are more common.

Masek
Fauna of this period is only known from one site. It includes some of the same animals that are found in Bed IV, notably the suids *Mesochoerus* and *Stylochoerus*, as well as *Phacochoerus* and a gerbil.

The Olduvai Remains

Ndutu

Living species of animals are found in the upper Ndutu Beds. The fauna from the Lower Ndutu Beds is virtually unknown except for a giant form of *Theropithecus*.

Naisiusiu

Fauna from one excavated site consists of extant species, including Burchell's zebra, living on the Serengeti Plain today.

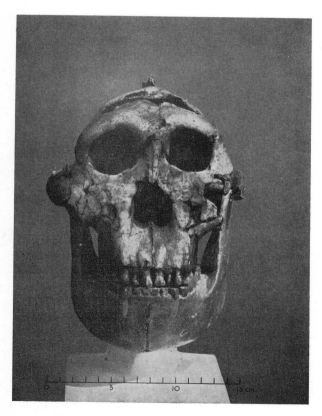

Fig. 42 The *Australopithecus boisei* skull articulated with a
modified Peninj mandible
Frontal view
*Courtesy of the late Dr L. S. B. Leakey and Bob Campbell,
Armand Denis Productions*

Dating An early potassium-argon date from a layer near the base of Bed I was given as 1·7 million years B.P. (Leakey, Evernden and Curtis, 1961); this age was disputed (von Koenigswald, Gentner and Lippolt, 1961) but subsequent work has produced a series of dates for the Olduvai stratigraphic units based on potassium-argon, geomagnetic polarity, fission track, C14, aminoacid racemization as well as sedimentation rates. Bed I (approx. 2·1–1·7 m.y. B.P.), Bed II (1·7–1·15 m.y. B.P.), Bed III (1·15–0·8 m.y. B.P.), Bed IV (0·8–0·6 m.y. B.P.), the Masek Beds (0·6–0·4 m.y. B.P.), the Ndutu Beds (approx. 400,000–32,000 years B.P.) and the Naisiusiu Beds (22,000–15,000 years B.P.) (Hay, 1975 in press; Hay, 1976).

Morphology 1. OLDUVAI HOMINID 5
This skull was recovered in 1959 from site FLK Bed I. The skull is almost complete with an intact upper dentition, only the mandible is lacking. It is probably male and belonged to a

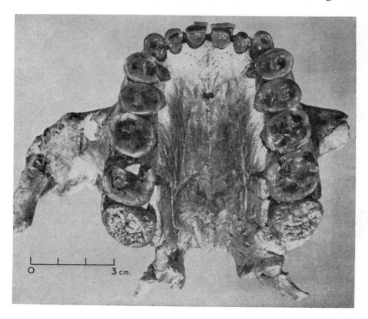

Fig. 43   The palate and dentition of *Australopithecus boisei*
Courtesy of the late Dr L. S. B. Leakey, photographed by R. Klomfass

young adult whose third molars had erupted, but had not come into wear; the sutures of the skull are still open. The skull has been fully described (Tobias, 1967). The cranial vault is low and the brow ridges are strongly marked; the facial skeleton and palate are large and the teeth robust. Anteroposterior dental disproportion is marked in that the incisors and canines are relatively small while the molars and premolars are large. The premolars show marked 'molarization' and the third molars are heavily wrinkled. A particularly striking feature of the skull is the great development around the cranium of muscular ridges such as the sagittal, occipital and supramastoid crests which were probably associated with a heavy mandible and powerful muscles of mastication. The mastoid processes are large and laterally prominent.

OLDUVAI HOMINID 20

This femoral fragment was recovered from site HWK as a surface find that is believed to have come from Upper Bed I or Lower Bed II.

The fragment consists of the upper end of the shaft, greater trochanter, lesser trochanter and neck of a robust left femur. The head of the bone is missing as well as small portions of both greater and lesser trochanters. The neck of the bone is flattened, the lesser trochanter posteriorly directed and the greater trochanter lacks lateral flare. The relative length of the neck of the bone is high by comparison with those of modern man and the great apes (Day, 1969).

2. OLDUVAI HOMINID 7

The remains described under this number come from site FLKNN Bed I, and comprise parts of a juvenile individual. It includes a damaged mandible with a partial dentition (third molar not yet erupted), a pair of parietals, an occipital fragment, fragments of both right and left petrous portions of the temporal bones, several other skull fragments and a group of hand bones (Leakey, 1960, 1961a and b). (Of the original twenty-one 'hand' bones six are non-hominid and one appears to be a vertebral fragment.) Of the fifteen left for consideration seven are clearly juvenile, six are of uncertain age and two are adult. It may be that the adult bones should be excluded from further consideration (Day, 1976). Of the 138

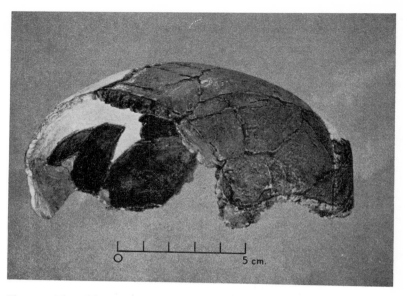

Fig. 44   The Olduvai Hominid 7 biparietal calvarial arch
Type specimen of *Homo habilis*
Reconstructed by Professor Tobias and A. R. Hughes
*Courtesy of the late Dr L. S. B. Leakey and Professor P. V. Tobias,
photographed by A. R. Hughes*

thirteen that are left one is clearly left sided, four are right sided and all are juvenile or possibly juvenile. Most economically at least two hands are involved.

The *mandible* was broken prior to fossilization and the right side of the body has been displaced medially. The body is very stout but its depth cannot be determined since the lower border and part of the symphysial region are missing. Both rami and their coronoid and condyloid processes are also absent. All of the permanent teeth are present in the mandible with the exception of the right second molar and both third molars. Sufficient bone is present distal to the left second molar to establish that the third molar is unerupted. The incisors are hominid in their general structure, but at least one of the premolars is elongated in that the mesiodistal length of its crown is greater than its buccolingual width. The second   **139**

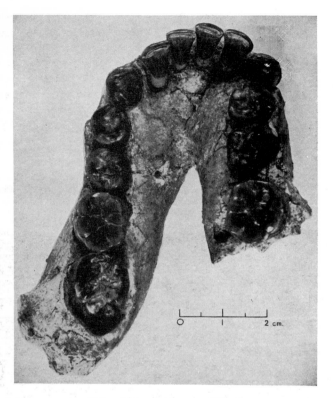

Fig. 45   The Olduvai Hominid 7 mandible
Type specimen of *Homo habilis*
Courtesy of the late Dr L. S. B. Leakey, photographed by
A. R. Hughes

molar crown is larger than the first on the left side and the
molar cusp pattern is basically dryopithecine.

The *parietals* are thin and immature and have no sign of a
sagittal crest or marked temporal lines. Restoration was
facilitated by the fortunate preservation of the entire coronal
and temporal borders of the left bone and the entire occipital
border of the right bone. Moreover the anterior part of the
sagittal border of the left parietal bone and the posterior part
of this border of the right parietal bone were preserved. Thus
although the bones have no point of contact, reconstruction

of the vault could proceed. The volume of the biparietal endocast was estimated by water displacement (see page 148) (Leakey, 1961a; Tobias, 1964b).

The *hand* was powerfully built, capable of strong finger flexion and possessed an opposable thumb; deductions based upon the morphology of the finger bones, the presence of a well defined saddle surface on the trapezium and a thumb terminal phalanx that is stout and broad (Napier, 1962; Day, 1976).

Fig. 46 The Olduvai Hominid 7 Hand. Reconstructed by M. H. Day and J. L. Scheuer

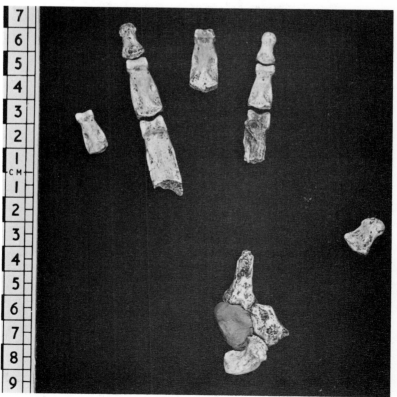

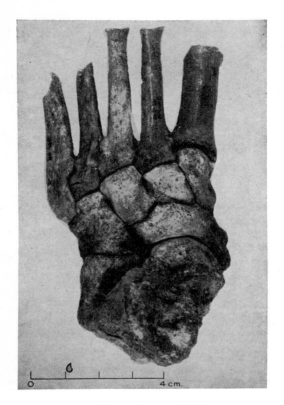

Fig. 47   The Olduvai (Hominid 8) foot
*Photographed by courtesy of the late Dr
L. S. B. Leakey*

OLDUVAI HOMINID 8
This specimen was recovered from site FLKNN, Bed I
(Leakey, 1960); all of the bones were found within the area of
one square foot. It comprises the bones of a foot, including all
the left tarsal and metatarsal bones, of an adult individual; all
the phalanges, the metatarsal heads and the posterior portion
of the calcaneum are missing. It is clear from examination of
this foot when articulated that it has most of the specializations
associated with the plantigrade propulsive feet of man (Day
and Napier, 1964). The presence of an articular facet between
the bases of the first and second metatarsals demonstrated   142

unequivocally the absence of hallucial divergence which characterizes non-human primate feet. The distal row of tarsal bones forms a well marked transverse arch; the ligamentous and muscular impressions upon the bones provide evidence for the static and dynamic support of the arches of the foot. In short the foot is non-prehensile and adapted for upright stance and bipedal gait.

OLDUVAI HOMINID 10

This specimen was recovered from site FLK (North) at the top of Bed I. It is the terminal phalanx of a right great toe. The bone is short, broad and flattened; its head bears a prominent tubercle for the support of the nail while its base has stout tubercles for the collateral ligaments. The long axis of the bone shows a valgus deviation and the shaft shows lateral torsion. It is believed that this terminal phalanx provides further evidence for an advanced form of bipedal gait in the early hominid forms which were evolving in East Africa during the Lower Pleistocene (Day and Napier, 1966; Day, 1967).

Fig. 48 The palate and upper dentition of Olduvai Hominid 13 Occlusal view *Photographed by courtesy of Dr M. D. Leakey*

0   2   4   6   8

143

OLDUVAI HOMINID 13
This skull was found at site MNK in Lower Bed II. It consists
of the greater part of the mandible including the complete
dentition, the greater part of the maxillae, most of the
occipital, the right parietal and temporal bones, parts of the
left parietal and temporal bones as well as frontal fragments
and many other bone fragments. It has been possible to
reconstruct the dental arcades and to place them into occlusion
and to attempt a reconstruction of the skull vault. The teeth
are elongated mesiodistally and narrowed labiolingually
(Leakey and Leakey, 1964).

Fig. 49   The Olduvai Hominid 13 mandible and lower dentition
Occlusal view
Photographed by courtesy of Dr M. D. Leakey

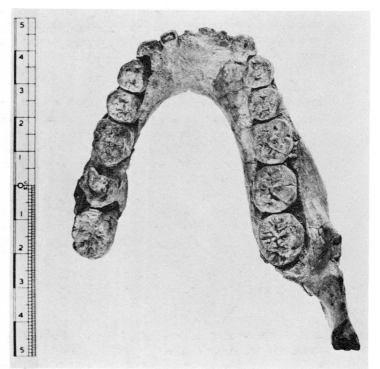

The Olduvai Remains

OLDUVAI HOMINID 24

This skull was recovered from Lower Bed I at site DK East. The specimen includes the greater part of the vault, much of the occipital bone, nearly all of the sphenoid, nearly all of the right temporal bone and a considerable part of the face and palate.

Unfortunately the skull was badly crushed while in the deposit and despite a remarkable reconstruction it retains a good deal of distortion. In particular the reconstructed vault remains flattened from above downwards and this has resulted in the occiput protruding further posteriorly than it should. The face is concave when viewed in profile with no nasal protrusion and the supraorbital ridge is prominent and continuous across a marked glabellar protrusion. From above, the vault outline shows considerable postorbital waisting, but this may appear exaggerated since the vault is much reconstructed and asymmetrical. As the earliest of the hominids recovered from Olduvai it has been described in some detail in the preliminary communication (Leakey, Clarke and Leakey, 1971).

OLDUVAI HOMINID 35

Early in 1960 a tibia and fibula were found on the same living floor as the Olduvai Hominid 5 skull at site FLK (Leakey, 1960). The upper ends of both bones are missing but the lower articular surfaces are intact allowing them to be fitted together. In a preliminary assessment the bones were described as being well adapted to bipedal gait at the ankle, but less well adapted at the knee. It was concluded that while this form was an habitual bipedal walker, the gait may have differed considerably from that of man (Davis, 1964). Subsequent examination of the upper end of the tibia has suggested an alternative explanation of part of the anatomy which may detract from the view that the leg was less well adapted for bipedalism at the knee (Day, 1976).

OLDUVAI HOMINID 48

A clavicle was recovered from site FLKNN in Bed I (Leakey, 1960) and remains unique as the only early hominid clavicle known from any site in the world, some 'clavicular' fragments

145

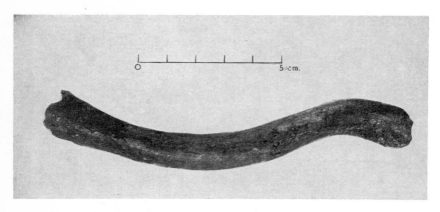

Fig. 50   The Olduvai (Hominid 48) clavicle
Superior view
*Photographed by courtesy of the late Dr L. S. B.
Leakey*

reported from the Transvaal having subsequently been discounted. As originally described it is said to have overall
similarities to that of man (Leakey, Tobias and Napier 1964).
Subsequent examination has done little to alter that view
(Day, 1976).

3. OLDUVAI HOMINID 9

This find consists of the greater part of a thick calvaria that
was recovered from site LLK, Upper Bed II. It lacks a face,
base and part of the vault. The frontal region is flattened, the
nuchal crest is prominent, the inion coincides with the
opisthocranion and the nuchal plane is flattened. The small
mastoid processes are surmounted by supramastoid crests
that are continuous with the nuchal crest and the brow ridges
are large and flaring (Leakey, 1961a).

OLDUVAI HOMINID 28

This femoral shaft and hip bone were recovered together *in*
situ from Bed IV at site WK. The femur lacks a head, greater
trochanter and the lower end. It is flattened throughout with
acute medial and lateral borders, it has a convexity of the
medial border and the narrowest point of the shaft is distally
placed. The hip bone is represented by the major parts of its   146

Fig. 51   The Olduvai Hominid 9 calvaria
Reconstructed by Professor P. V. Tobias and A. R. Hughes
Courtesy of the late Dr L. S. B. Leakey and Professor P. V. Tobias,
photographed by A. R. Hughes

iliac and ischial portions, the pubic portion is missing. The
bone is stoutly constructed with prominent muscular markings and a large acetabulum indicating a large femoral head.
The iliac pillar is massive and the ischium medially rotated
(Day, 1971).

OLDUVAI HOMINID 36

During the excavation of site SC, Upper Bed II, an almost
complete right ulna was recovered. The most striking features
of the bone are its general robustness, the strength of its
muscular markings and the degree of its curvature in the
anteroposterior plane throughout its length. When viewed
posteriorly the bone shows a typical ulnar sigmoid. The
trochlear notch is unusual in several respects; in particular
it has a smooth vertical ridge or keel dividing the articular
surface into a small medial portion and a large lateral portion
while the gape of the trochlear notch is small in relation to the
overall size of the bone.   147

Dimensions   I. OLDUVAI HOMINID 5
Tobias, (1963, 1967)

Maximum Length (Glabella-Opisthocranion) 173·0
Maximum Breadth (Supramastoid)   139·5
Maximum Height (Basion to a point vertically above it
   in the coronal plane)   Left of sagittal crest 92·0
                            Right of sagittal crest 90·5
Cranial Capacity   530 cc

2. OLDUVAI HOMINID 7
SKULL
Estimated Cranial Capacity 673·5–680·8 cc (Central values)
Range of estimates 642·7–723·6 cc (Tobias, 1964b, 1965c,
1966b; Holloway, 1965, 1966). New estimate 657 cc
(Tobias, 1968) and again more recently 687 cc (Tobias,
1971).

MANDIBLE
Dimensions not available.

TEETH
Tobias and von Koenigswald (1964)

		Lower Teeth (Crown Dimensions)					
	I_1	I_2	C	PM_1	PM_2	M_1	M_2	M_3
Left l	★	★	★	9·6	10·3	14·3	15·6	—
side b	★	★	★	10·3	10·7	12·2	13·5	—
Right l	★	★	★	9·9	11·1	14·3	—	—
side b	★	★	★	10·1	10·7	12·4	—	—

★ Unpublished

POSTCRANIAL BONES
Hand

Specimen	Digit	Inter-artic length	Width (med./lat.)	Depth (ant./post.)
D	II	18·5	9·8	4·3
F	III	25·1	11·8	5·0
E	IV	24·9	11·4	4·9
G	V	17·8	9·0	3·9

The Olduvai Remains

OLDUVAI HOMINID 10
Max. Length 17·5   Max. Breadth 18·3
Length/Breadth Index 104·5   Valgus Deviation 2°
Lateral Torsion 13°

OLDUVAI HOMINID 13
TEETH
Tobias and Von Koenigswald (1964)

| | | Upper and Lower Teeth (Crown Dimensions) | | | | |
		PM1	PM2	M1	M2	M3
Upper teeth	l	8·3	8·7	12·5	12·8	12·0
	b	11·4	11·5	12·6	13·8	13·1
Lower teeth	l	—	8·9	12·7	13·8	14·6
	b	—	9·8	11·6	12·2	12·5

Cranial Capacity (reconstructed skull) 650 cc (Holloway, 1973)

OLDUVAI HOMINID 24
SKULL
Leakey, Clarke and Leakey (1971)
Maximum Length (Glabella to inion chord) 147*
Maximum Breadth (Bimastoid)   122*
Cranial Capacity   590 cc*
                                                        (Holloway, 1973)

* Skull reconstructed with considerable residual distortion.

3. OLDUVAI HOMINID 9
SKULL
Length 209   Breadth 150   Cranial Index c. 72

OLDUVAI HOMINID 28
Day (1971)
FEMUR
Length (reconstructed)   456
Mid-shaft Diam. (sagittal)   24·7
                   (transverse) 32·7
Platymeric Index   62·3
Pilastric Index   75·5

The Olduvai Remains

Affinities   1. In the original description of the Olduvai Hominid 5 skull it was stated that, while the new skull belonged patently to the sub-family Australopithecinae, it differed from both *Australopithecus* and *Paranthropus* 'much more than these two genera differ from each other' (Leakey, 1959). On these grounds the new genus *Zinjanthropus* was created; the specific name *boisei* was given after a benefactor. Robinson (1960) believed that the cranial and dental characters of this skull are typically those of *Paranthropus* and proposed that *Zinjanthropus* be included in this genus (*Paranthropus boisei*). Following the discovery of further hominid remains in Bed I (*Homo habilis*), Leakey, Tobias and Napier (1964) recognized the genus *Australopithecus* as having three sub-genera within the family Hominidae, i.e. *Australopithecus, Paranthropus* and *Zinjanthropus*. Thus '*Zinjanthropus*' was accepted as an East African australopithecine, sub-generically distinct from the South African forms. Later Tobias (1967) formally sank the sub-genus '*Zinjanthropus*' and retained Hominid 5 within the genus and species *Australopithecus boisei*.
Olduvai Hominid 20 has been attributed to *Australopithecus cf. boisei* until further material is available for comparison.

2. Following the discovery and preliminary evaluation of this material it was considered that this form was not an australopithecine and that sufficient evidence was available on which to create a new species of the genus *Homo* in the light of a revision of the diagnosis of the genus. This revision and a definition of the new species was given by Leakey, Tobias and Napier (1964) and Olduvai Hominid 7 was designated the type specimen (juvenile mandible, skull bones and hand bones). Hominids 4, 6, 8, 13 and 48 (formerly included in Hominid 8) were declared paratypes of the new species which was named *Homo habilis*.
A debate ensued regarding the taxonomic status of *Homo habilis* during which Leakey, Tobias and Napier vigorously defended the new species (Tobias, 1965a, 1965b, 1966a; Leakey, 1964a and b; Napier, 1965) against several others who did not believe the new species to be a valid taxon or who felt that the evidence presented for this step was inadequate (Robinson, 1965a and b, 1966; Campbell, 1964; Oakley, 1964;   150

Le Gros Clark, 1964). Subsequently the intensity of the debate declined but it would be premature even now to say that it is resolved. Recent reassessment of the Olduvai post-cranial material in the light of that from East Rudolf has led to the suggestion that the Olduvai material may well be a mixture of hominine and australopithecine bones (Day, 1976).

3. Of this group of material Hominid 9, a calvaria, was at first likened to the 'Pithecanthropines' although some resemblance to later forms was suggested (Leakey, 1961a). Without further published detailed study it has widely become regarded as an example of *Homo erectus*, and has been so attributed (Leakey, 1971a).
On the basis of the similarities between the Hominid 28 femur and those from Peking both the Hominid 28 femur and its associated pelvic fragment have been attributed to *Homo erectus* (Day, 1971). Doubts have been cast on this attribution on the grounds that the pelvic specimen is 'merely aberrant', and that the anatomical association of the femoral and the pelvic fragment has been assumed (Brain, Vrba and Robinson, 1974). Since the two specimens were recovered from the same level within one square metre of the deposit (Leakey, 1971b), and a second specimen with almost identical features has been recovered from East Rudolf (Leakey, 1976) the doubts expressed may well be assuaged.

No.	Attribution	Stratum; Locality	Remains	Date	Associated Industry
O.H. 1	*H. sapiens*	Intrusive burial, Bed II: RK	Complete skeleton	1913	Nil
O.H. 2	cf. *H. erectus*	Surface, Bed IV, MNK	Two vault fragments	1935	Inferred Acheulean
O.H. 3	cf. *Australopithecus*	*In situ*, Upper Bed II; BK	One deciduous canine One deciduous molar	1955	Developed Oldowan
O.H. 4	*H. habilis*	*In situ*, Lower Bed I and surface; MK	One molar, two broken teeth	1959	Inferred Oldowan
O.H. 5*	*A. boisei*	Surface, Bed 1; FLK	Almost complete cranium	1959/60	Oldowan
O.H. 6	*H. habilis*	Surface, Middle Bed I	Two teeth, skull fragments	1959/60	Inferred Oldowan
O.H. 7*	*H. habilis* (type)	*In situ*, Middle Bed I; FLK NN	Mandible, parietals, hand bones	1960	Oldowan
O.H. 8*	*H. habilis*	*In situ*, Middle Bed I; FLK NN	Foot	1960	Oldowan
O.H. 9*	*H. erectus*	Surface, Upper Bed II; LLK	Calvaria	1960	Nil
O.H. 10*	?	Upper Bed I; FLK North	Terminal phalanx great toe	1961	Oldowan
O.H. 11	*Homo sp.*	Surface, possibly from Lower Ndutu Beds; DK	Palate and maxillary arch	1962	Nil
O.H. 12	*H. erectus*	Surface, Bed IV; VEK	Palate, maxillary arch and cranial fragments	1962	Inferred Acheulean
O.H. 13*	*H. habilis*	*In situ*, Lower Middle Bed II; MNK	Mandible, parietals, occipital, maxilla and many cranial fragments	1963	Oldowan
O.H. 14	?	Surface, inferred Lower Middle Bed II; MNK		1963	?Inferred Oldowan
O.H. 15	cf. *H. erectus*	*In situ*, Lower Middle Bed II; MNK	One worn canine, two molars	1963	Indet.
O.H. 16	?	*In situ*, Base Bed II; FLK Maiko Gully	Many skull fragments, upper and lower dentition	1963	Inferred Oldowan
O.H. 17 O.H. 18 O.H. 19	?	Surface; FLK Maiko Gully	Part of a deciduous molar	1963	Nil
O.H. 20*	*A. cf. boisei*	Surface, inferred Lower Bed II/Upper Bed I; HWK	Neck of femur	1959	Inferred Developed Oldowan A or Oldowan

No.	Attribution	Stratum; Locality	Remains	Date	Associated Industry
O.H. 21	cf. H. habilis	Surface, Bed I; FLK North	Upper molar	1968	Nil
O.H. 22	cf. H. erectus	Surface, indeterminate gravel? Bed IV (VEK/MNK)	Right half of mandible	1968	Nil
O.H. 23	cf. H. erectus	In situ, Masek Beds; FLK	Left mandibular fragment	1968	Acheulean
O.H. 24*	H. habilis	Surface, Lower Bed I; DK East	Cranium with some teeth	1968	Inferred Oldowan
O.H. 25	Indet.	Surface, Bed IV; Geol. Loc. 54	Parietal fragment (juv.)	1968	Inferred Acheulean
O.H. 26	? A. boisei	Surface, Upper Bed I Lower II; FLK West	Large unworn third molar	1969	Nil
O.H. 27	cf. H. habilis	Upper Bed I; Surface HWK	Unworn third molar	1969	Inferred Developed Oldowan A
O.H. 28*	H. erectus	In situ, Bed IV; WK	Left innominate, left femoral shaft	1970	Acheulean
O.H. 29	cf. H. erectus	In situ, Bed III; JK West	Molar fragment, 2 incisors and phalanx	1969	Acheulean
O.H. 30	Indet.	Surface, Lower Bed II; Maiko Gully North	Teeth, permanent and deciduous, cranial fragments	1969	Inferred Oldowan
O.H. 31	Indet.	Surface, Upper Bed I; HWK East	Part of a molar tooth	1969	Nil
O.H. 32	Indet.	Surface, Middle Bed II; MNK	Part of a molar tooth	1969	Nil
O.H. 33	Indet.	Surface, Bed I; FLK NN	Thin skull fragments	1969	Nil
O.H. 34	Homo sp.	In situ, Bed III; JK West	Femur and part tibia shaft	1962	Acheulean
O.H. 35*	H. habilis	In situ, Middle Bed II; FLK	Tibia and fibula	1960	Oldowan
O.H. 36*	H. erectus	Bed II (Tuff IIID); SC	Almost complete ulna	1970–71	?
O.H. 37	?	Surface, FLK; ex-Bed II	Left half mandible	1971	?
O.H. 38					
O.H. 39	?	Surface, ex Upper Bed I; (Matrix) HWK EE	Partial permanent and deciduous dentition	1972	Nil
O.H. 40	?	Surface, lower Bed II; FLK South	Broken molar (possibly dm^2)	1972	Nil

No.	Attribution	Stratum; Locality	Remains	Date	Associated Industry
O.H. 41	H. habilis	In situ, lower Bed II; HWK EE	Left M^1 or M^2	1972	Oldowan
O.H. 42	?	Surface, upper Bed I; HWK EE	Broken premolar	1972	Nil
O.H. 43	?	In situ, Bed I; FLK NN	2 left metatarsals	1960	Oldowan
O.H. 44	H. habilis	Surface, Bed I; FLK at site of O.H. 5	Right M^1	1970	Nil
O.H. 45	H. habilis	In situ, Bed I; FLK NN	Left germ M^1	1960	Nil
O.H. 46	cf. A. boisei	In situ, Bed I; FLK NN	Broken crown of molar or premolar	1960	Nil
O.H. 47 O.H. 48*	? H. habilis	In situ, Middle Bed I; FLK NN	Clavicle	1960	Oldowan
O.H. 49	? H. habilis	In situ, Middle Bed I; FLK NN	Part radius shaft	1960	Oldowan
O.H. 50	?	In situ, Middle Bed I; FLK NN	Part rib shaft	1960	Oldowan
O.H. 51	cf. H. erectus	Surface, Bed III; GTC	Part left side of mandible with P_4, M_1	1974	Nil
O.H. 52	cf. H. habilis	Surface, lower Bed I; DK East	Incomplete left temporal	1969	Nil
O.H. 53	?	In situ, Middle Bed II; SHK	Shaft of right femur	1957	Developed Oldowan
O.H. 54	?	Surface, Bed I; Maiko Gully	Half lower M_1 or M_2	1969	Nil
O.H. 55	H. habilis	Surface, Bed I; FLK South	Crown of molar, weathered	1976	Nil

The Olduvai Remains

Originals Property of the Government of Tanzania, East Africa; to be housed in the National Museum, Dar-es-Salaam; at present being held at the National Museum of Kenya, Nairobi.

Casts Dr M. D. Leakey, c/o National Museum of Kenya, Nairobi.

References LEAKEY, L. S. B. 1959
A new fossil skull from Olduvai. *Nature 201,* 967-970.
LEAKEY, L. S. B. 1960
Recent discoveries at Olduvai Gorge. *Nature 188,* 1050-1052.
ROBINSON, J. T. 1960
The affinities of the new Olduvai australopithecine. *Nature 186,* 456-458.
KOENIGSWALD, G. H. R. VON, GENTNER, W. and LIPPOLT, H. J. 1961
Age of the basalt flow at Olduvai, East Africa. *Nature 192,* 720-721.
LEAKEY, L. S. B. 1961a
New finds at Olduvai Gorge. *Nature 189,* 649-650.
LEAKEY, L. S. B. 1961b
The juvenile mandible from Olduvai. *Nature 191,* 417-418.
LEAKEY, L. S. B., EVERNDEN, J. F., and CURTIS, G. H. 1961
Age of Bed I, Olduvai Gorge, Tanganyika. *Nature 191,* 478-479.
NAPIER, J. R. 1962
Fossil hand bones from Olduvai Gorge. *Nature 196,* 409-411.
HAY, R. L. 1963
Stratigraphy of Bed I through IV, Olduvai Gorge, Tanganyika. *Science N.Y. 139,* 829-833.
HEBERER, G. 1963
Über einen neuen archanthropinen Typus aus der Oldoway Schlucht. *Z. Morph. Anthr. 53,* 171-177.
TOBIAS, P. V. 1963
Cranial capacity of Zinjanthropus and other australopithecines. *Nature 197,* 743-746.
CAMPBELL, B. G. 1964
Just another man-ape? *Discovery* (June issue).
CLARK, W. E. LE GROS. 1964
Letter. *Discovery* (July issue).
DAVIS, P. R. 1964
Hominid fossil from Bed I, Olduvai Gorge, Tanganyika. A tibia and fibula. *Nature 201,* 967-970.
DAY, M. H. and NAPIER, J. R. 1964
Hominid fossils from Bed I, Olduvai Gorge, Tanganyika. Fossil foot bones. *Nature 201,* 967-970.

LEAKEY, L. S. B. 1964a
Letter. *Discovery* (August issue).
LEAKEY, L. S. B. 1964b
Letter. *Discovery* (October issue).
LEAKEY, L. S. B. and LEAKEY, M. D. 1964
Recent discoveries of fossil hominids in Tanganyika: at Olduvai and near Lake Natron. *Nature 202*, 5–7.
LEAKEY, L. S. B., TOBIAS, P. V. and NAPIER, J. R. 1964
A new species of the genus *Homo* from Olduvai Gorge. *Nature 202*, 7–9.
NAPIER, J. R. 1964
Five steps to man. *Discovery* (June issue).
OAKLEY, K. P. 1964
Letter. *Discovery* (September issue).
TOBIAS, P. V. 1964a
Letter. *Discovery* (September issue).
TOBIAS, P. V. 1964b
The Olduvai Bed I hominine with special reference to its cranial capacity. *Nature 202*, 3–4.
TOBIAS, P. V. and KOENIGSWALD, G. H. R. VON 1964
Comparison between the Olduvai hominines and those of Java and some implications for hominid phylogeny. *Nature 204*, 515–518.
HOLLOWAY, R. L. 1965
Cranial capacity of the hominine from Olduvai, Bed I. *Nature 208*, 205–206.
LEAKEY, L. S. B. 1965
Olduvai Gorge 1951–1961, Vol. 1. Cambridge: Cambridge University Press.
NAPIER, J. R. 1965
Curr. Anthrop. Comment. *Curr. Anthrop. 6*, 402–403.
ROBINSON, J. T. 1965a
Homo 'habilis' and the australopithecines. *Nature 205*, 121–124.
ROBINSON, J. T. 1965b
Curr. Anthrop. Comment. *Curr. Anthrop. 6*, 403–406.
TOBIAS, P. V. 1965a
New discoveries in Tanganyika: their bearing on hominid evolution. *Curr. Anthrop. 6*, 391–399.
TOBIAS, P. V. 1965b
New discoveries in Tanganyika: their bearing on hominid evolution. Curr. Anthrop. comment. *Curr. Anthrop. 6*, 406–411.
TOBIAS, P. V. 1965c
Cranial capacity of the hominine from Olduvai, Bed I. *Nature 208*, 206.
DAY, M. H. and NAPIER, J. R. 1966
A hominid toe bone from Bed I, Olduvai Gorge, Tanzania. *Nature 211*, 929–930.

LEAKEY, M. D. 1966
A review of the Oldowan culture from Olduvai Gorge, Tanzania. *Nature 210*, 462-466.
ROBINSON, J. T. 1966
The distinctiveness of *Homo habilis*. *Nature 209*, 957-960.
TOBIAS, P. V. 1966a
The distinctiveness of *Homo habilis*. *Nature 209*, 953-957.
TOBIAS, P. V. 1966b
Cranial capacity of the Olduvai Bed I hominine. *Nature 210*, 1108-1109.
DAY, M. H. 1967
Olduvai Hominid 10: a multivariate analysis. *Nature 215*, 323-324.
TOBIAS, P. V. 1967
Olduvai Gorge, Vol. 2. Cambridge: Cambridge University Press.
TOBIAS, P. V. 1968
Cranial capacity in anthropoid apes, *Australopithecus* and *Homo habilis*, with comments on skewed samples. *S. Afr. J. Sci. 64*, 81-91.
DAY, M. H. 1969
Femoral fragment of a robust australopithecine from Olduvai Gorge, Tanzania. *Nature 221*, 230-233.
DAY, M. H. 1971
Postcranial remains of *Homo erectus* from Bed IV, Olduvai Gorge, Tanzania. *Nature 232*, 383-387.
LEAKEY, M. D. 1971a
Olduvai Gorge, Vol. 3. Cambridge: Cambridge University Press.
LEAKEY, M. D. 1971b
Discovery of postcranial remains of *Homo erectus* and associated artefacts in Bed IV at Olduvai Gorge, Tanzania. *Nature 232*, 380-383.
LEAKEY, M. D., CLARKE, R. J. and LEAKEY, L. S. B. 1971
New hominid skull from Bed I, Olduvai Gorge, Tanzania. *Nature 232*, 308-312.
TOBIAS, P. V. 1971
The brain in hominid evolution. New York: Columbia University Press.
ROBINSON, J. T. 1972
Early hominid posture and locomotion. Chicago: Chicago University Press.
HOLLOWAY, R. L. 1973
New endocranial values for the East African early hominids. *Nature 243*, 97-99.
KLEINDIENST, M. R. 1973
Excavations at Site J.K.2, Olduvai Gorge, Tanzania, 1961-1962: the geological setting. *Quaternaria 17*, 145-208.
BRAIN, C. K., VRBA, E. S. and ROBINSON, J. T. 1974
A new hominid innominate bone from Swartkrans. *Ann. Transv. Mus. 29*, 55-63.

The Olduvai Remains

HAY, R. L. 1975
Olduvai Gorge: stratigraphy, age and environments. In *Geological Background to Fossil Man*. Ed. W. W. Bishop (in press).
DAY, M. H. 1976
Hominid postcranial material from Bed I, Olduvai Gorge. In *Perspectives on Human Evolution Vol. III: Human Origins*. 363-374. Eds. G. L. Isaac, and E. R. McCown, Menlo Park, Cal.: W. A. Benjamin Inc.
HAY, R. L. 1976
Geology of the Olduvai Gorge: a study of the sedimentation in a semiarid basin. Berkeley: University of California Press.
LEAKEY, R. E. F. and FINDLATER, I. 1976
New hominid fossils from the Koobi Fora Formation in Northern Kenya. *Nature 261*, 574-6.

The Laetolil Remains

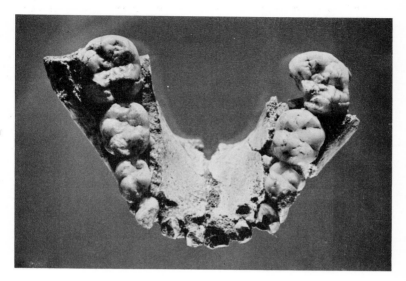

Fig. 52   The Laetolil Hominid 2 juvenile mandible and dentition
Occlusal view
Courtesy of Dr M. D. Leakey

Synonyms and other names	Cf. *Homo* sp. (Leakey, M. D., Hay, Curtis, Drake, Jackes and White, 1976)
Site	Laetolil, southern Serengeti Plains, northern Tanzania, 20 to 30 miles from the camp-site at Olduvai Gorge.
Found by	Dr Mary Leakey and her co-workers, 1974–1975.
Geology	Recent work on the geology of the Laetolil area has shown that the Laetolil Beds (formerly referred to variously as the Garusi or Vogel River series) proved to be far thicker than previously recognized. The lavas and agglomerates noted by Kent (1941) overlie an irregular surface deeply eroded into the Laetolil Beds. The 130 m. section of the Laetolil Beds is divisible into an upper half consisting of aeolian tuff and a lower half consisting of interbedded ash-fall and aeolian tuff with minor conglomerate and breccia. All of the hominid

159

remains and most of the faunal remains come from the uppermost 30 m. of the aeolian tuffs. Beds I and II, Olduvai Gorge (*q.v.*) are represented at Laetolil and overlie the Laetolil Beds (Hay, in Leakey *et al.*, 1976).

Associated finds   No stone tools have been recovered from this site up to the present.

Fossils have been recovered in the area on several occasions, the largest collection, including a hominid fragment, being made by Kohl-Larsen in 1938–1939. Later, discrepancies were noted in the fauna by Dietrich (1942) and Maglio (1969) who both suggested two faunal assemblages of differing ages. In the recent work fossils from the Laetolil Beds were noted as being distinctively coloured and of a chalky texture. Representatives of vertebrate groups assigned to these Beds include bovids, lagomorphs, giraffids, rhinocerotids, equids, suids, proboscidians, rodents and carnivores. Some genera have now been excluded from former lists given by Dietrich (*op. cit.*) and Hopwood (in Leakey, 1951); these include *Theropithecus*, *Tragelaphus*, *Equus* and *Phacochoerus*.

Dating   Stratigraphically the Laetolil Beds underlie Bed I and Bed II, Olduvai Gorge, thus it is likely that they must be as old as 2 m.y. B.P. at the very least. Potassium–argon dating of a series of suitable deposits from the site bracket the hominid remains and give dates that range from 3·41–3·7 m.y. B.P. The average date given for the hominid horizon is 3·59 m.y. B.P. on the basis of both conventional potassium–argon and Ar^{40}/Ar^{39} ratio methods.

Morphology   Thirteen new hominid specimens have been recovered from the Laetolil site during 1974 and 1975; the remains include mandibles, maxillae and isolated teeth.

L.H.2

This specimen consists of the body of the mandible of an infant and contains part of both the permanent and deciduous dentitions. The mandibular symphysis is incompletely fused in the specimen and both of the rami are missing. The symphysial region shows a concave *planum alveolare* and an incipient superior transverse torus.

The deciduous incisors are unknown but the right deciduous
canine is present as a damaged but sharp conical tooth; the
lower first deciduous molar has four or five main cusps and a
lingually directed anterior fovea; the lower second deciduous
molar has a sixth cusp. The lower first permanent molar is
square with five cusps present.

L.H.4

This specimen consists of the body and part of the dentition
of an adult mandible. The rami are missing. Much of the
anterior dentition has been lost post-mortem except for the
right lateral incisor which appears to have been lost during
life. There seems to be evidence of periodontal disease in
relation to this tooth and several others. The teeth appear to
have been arranged in an evenly rounded dental arcade and
there is development of both superior and inferior transverse
tori. The basal contour of the body of the mandible is wide
with marked lateral eversion posteriorly. The lower first
premolars have large buccal cusps and weak lingual cusps

Fig. 53   The Laetolil Hominid 4 adult mandible and dentition
          Occlusal view
          Courtesy of Dr M. D. Leakey

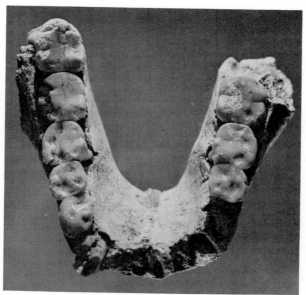

The Laetolil Remains

placed obliquely giving the tooth an irregular shape; the lower second premolars are square with buccal and lingual cusps and moderate talonids. The lower molars increase in size progressively from the first to the third and show a square occlusal outline with five cusps and no trace of a sixth cusp. The fissure pattern is Y5.

Dimensions   Dimensions to be found in a descriptive publication.

Affinities   In the preliminary description it has been suggested that the new Laetolil specimens represent one group of hominids only and that the variation shown in the sample is either individual or sexual in origin. Some similarities to the South African gracile *Australopithecus africanus* material has been noted as well as some to specimens attributed to the genus *Homo* from East Rudolf. Marked differences are suggested between the new Laetolil material and robust australopithecine specimens from both East and South Africa. In addition some similarities are claimed for the Laetolil hominids to specimens of *Homo erectus* from Asia as well as to some of the new material from Hadar, Ethiopia (*q.v.*).

Originals   To be housed in the National Museum of Tanzania, Dar-es-Salaam; at present being studied at the National Museum of Kenya, Nairobi.

Casts   Not available at present.

References   KENT, P. E. 1941
The recent history and Pleistocene deposits of the plateau north of Lake Eyasi, Tanganyika. Geol. Mag. Lond. 78, 173-184.
DIETRICH, W. O. 1942
Altestquartäre Saügetiere aus der südleichen Serengeti Deutsche-Ostafrika. Palaeontographica 94A, 43-133.
HOPWOOD, A. T. 1951
The Olduvai Fauna. In L. S. B. Leakey, Olduvai Gorge, 20-24. Cambridge: Cambridge University Press.
MAGLIO, V. J. 1969
The status of the East African elephant 'Archidiskodon exoptatus' Dietrich 1942. Breviosa 336, 1-25.
LEAKEY, M. D., HAY, R. L., CURTIS, G. H., DRAKE, R. E., JACKES, M. K. and WHITE, T. D. 1976
Fossil hominids from the Laetolil Beds. Nature, 262, 460-66.

The Peninj Mandible

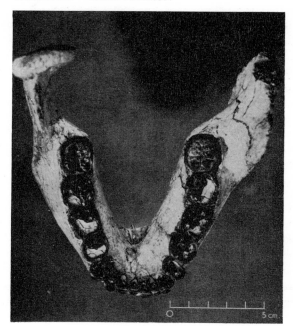

Fig. 54   The mandible from Peninj, Tanzania
*Courtesy of the late Dr L. S. B. Leakey, photo-
graphed by L. P. Morley*

*Synonyms
and other names*   *Australopithecus* (*Zinjanthropus*) *boisei* (Leakey and Leakey, 1964); *A. cf. boisei* (Tobias, 1968); The Natron mandible

Site   Peninj; a locality to the west of Lake Natron some 50 miles north-east of Olduvai Gorge, Tanzania.

Found by   Mr Kamoya Kimeu during an exploratory expedition led by R. E. F. Leakey and G. L. Isaac; January, 1964.

Geology   The Peninj group consists of an upper Moinink Formation and a lower Humbu Formation of Middle Pleistocene sediments including tuffs, tuffaceous shales, clays, sands and a locally interbedded olivine basalt flow. The detailed geology has been given by Isaac (1965, 1967). The mandible was found in a partially slumped block of sandstone which could be matched to a stratum outcropping in a nearby cliff.

163

The Peninj Mandible

Associated finds During the course of excavation some 120 artefacts were collected from the deltaic and alluvial facies of the Humbu Formation. These consist of cores, choppers and flakes as well as cleavers and some handaxes. In addition two archaeological sites were found that produced crude Lower Acheulean handaxe-like bifaces and cleavers.

An assemblage of mammalian fossils was collected including *Equidae, Rhinocerotidae, Cercopithecidae, Giraffidae* (including *Libytherium*), *Carnivora, Bovidae, Suidae* and *Elephantidae*. Both the bovid and suid faunal assemblages indicate a Middle Pleistocene age and are consistent with correlation of the Humbu Formation with Upper Bed II, Olduvai Gorge (Isaac, 1967).

Dating Faunal correlation has suggested a Middle Pleistocene age for the Humbu Formation but geophysical data reported recently are a little confusing. Two chronometric schemes have been proposed based on potassium-argon estimations and palaeomagnetic data. The first scheme proposes an age of about 1·9 m.y. B.P. and the normal polarity due to the Olduvai Event; the second scheme proposes an age of about 1 m.y. B.P. and the normal polarity due to the Jaramillo Event. Taken as a whole the evidence is said to indicate an age of the order of 1·5 m.y. B.P. for the site (Isaacs [sic] and Curtis, 1974).

Morphology The specimen consists of an almost perfect mandible containing the complete adult dentition. On the left, the condylar process is missing, and on the right the angle of the jaw is broken away. The body of the mandible is robust and deep, the symphysial region recedes and the extramolar sulcus is broad. Internally there is a postincisive *planum*, marked genial tubercles and fossae as well as a prominent *crista pharyngea*.

The teeth show anteroposterior disproportion in that the incisors and canines are tiny by comparison with the premolars and molars. The premolars are themselves highly molarified and worn almost flat. The third molars are erupted into wear and show a dryopithecine cusp and fissure pattern with secondary enamel wrinkling. The jaw shows multiple mental foramina on the right.

The Peninj Mandible

Dimensions   TEETH
(*Tobias* 1968)

		Lower Permanent Teeth (Crown Dimensions)						
		I_1	I_2	C	PM_1	PM_2	M_1	M_2	M_3
Left side	l	5·5	5·9	7·3	9·2	14·5	16·4	17·4	18·8
	b	6·1	6·4	8·1	13·5	15·2	15·3	16·2	16·3
Right side	l	5·6	6·2	7·4	9·9	12·7*	16·1	17·2	18·7
	b	6·1	6·5	8·3	13·5	14·8	15·5	16·1	15·7

* Damaged

Affinities   When first reported there was no hesitation in designating this jaw as 'an unmistakable australopithecine' (Leakey and Leakey, 1964), subsequently this opinion has been underlined by Tobias (1968) who placed the jaw in the genus and species *Australopithecus boisei* distinguishing it from *A. robustus* on a number of dental features.

Original   Property of the Government of Tanzania to be housed in the National Museum of Tanzania, Dar-es-Salaam; at present being studied at the National Museum of Kenya, Nairobi.

Casts   Dr M. D. Leakey, c/o National Museum of Kenya, P.O. Box 40658, Nairobi Kenya.

References   LEAKEY, L. S. B., and LEAKEY, M. D. 1964
Recent discoveries of fossil hominids in Tanganyika: at Olduvai and near Lake Natron. *Nature 202*, 5-7.
ISAAC, G. L. 1965
The stratigraphy of the Peninj Beds and the provenance of the Natron australopithecine mandible. *Quaternaria 7*, 101-130.
ISAAC, G. L. 1967
The stratigraphy of the Peninj Group—Early Middle Pleistocene formations west of Lake Natron, Tanzania. In *Background to evolution in Africa*, 229-257. Eds. W. W. Bishop and J. D. Clark. Chicago: Chicago University Press.
TOBIAS, P. V. 1968
The early hominid remains from Tanganyika: *Australopithecus* and *Homo. VIIme Congrès Internationale des Sciences Anthropologiques et Ethnologiques III*, 333-341.
ISAACS [*sic*], G. L. and CURTIS, G. H. 1974
Age of early Acheulean industries from the Peninj Group, Tanzania. *Nature 249*, 624-627.

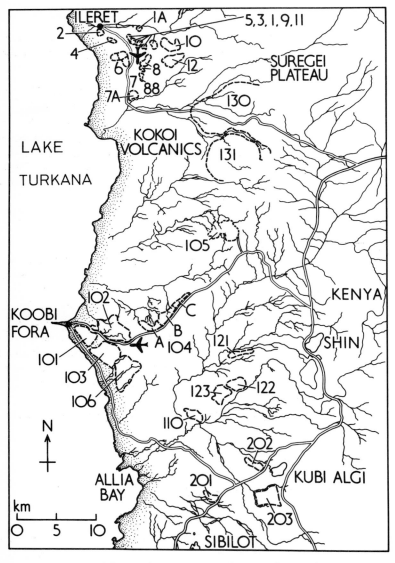

Fig. 55 A map of the Koobi Fora region showing the numbered areas in which finds have been made

The Koobi Fora Remains
(formerly East Rudolf Hominids)

Synonyms and other names	*Australopithecus cf. boisei* (Leakey, 1970); *Australopithecus sp.* (Leakey, 1971); *Homo sp.* (Leakey, 1971); *Homo ergaster* (Groves and Mazák, 1975)
Site	In north-east Kenya to the east of Lake Turkana. The area was formerly known as East Rudolf but is now called the Koobi Fora region. The research area extends from Ileret in the north to Allia Bay in the south, and from the shores of the lake in the west to the Miocene volcanic hills in the east; a total of 1,000 sq. kms. The research area has been sub-divided into numbered areas each corresponding to an area of exposure of fossiliferous sediments.
Found by	The East Rudolf Research Project led by R. E. F. Leakey and G. L. Isaac.
Geology	The basin of Lake Turkana appears to be of great antiquity and basement rocks are exposed only in the extreme north-east portion of the study area. On these there rests a long succession of Miocene to early Pliocene volcanics and sediments, some of which are fossiliferous. Downwarping along the axis now occupied by the lake led to the deposition of a long series of late Pliocene to early Pleistocene sedimentary beds which are interspersed with volcanic tuffs. At the basin margins these sediments lap against unwarped hills formed by rocks of the earlier volcanic series.

The stratigraphic terminology of the Plio-Pleistocene sediments is as follows:

> Guomde Formation
> disconformity
> Koobi Fora Formation { Upper Member/Ileret Member
>   { Lower Member
> Kubi Algi Formation

The principal hominid-bearing formation is the Koobi Fora Formation which has been subdivided into an Upper Member and a Lower Member separated by the KBS tuff. The Upper Member is termed the Ileret Member at Ileret. The Guomde **167**

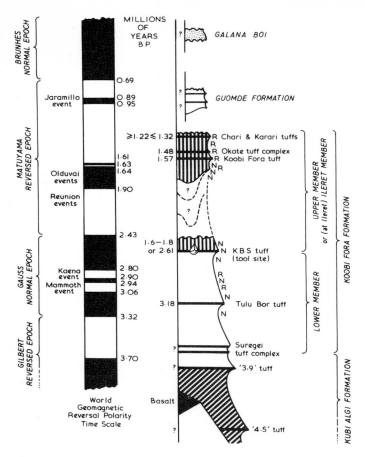

Fig. 56   The stratigraphic column of the Koobi Fora region compared with the world geomagnetic reversal polarity time scale
After Fitch, Findlater, Watkins and Miller, 1974

Formation overlies the Koobi Fora Formation and the Kubi Algi Formation underlies it.

The geological structure of the area and the history of sedimentation has been extensively determined by the development of the East African Rift system. The hominid-bearing fossiliferous sediments are situated on a shelf or block to the east of the trough which contains Lake Turkana. The 168 block is bounded on the other side by a complex fault zone which links the main Kenya Rift with the Ethiopian Rift. The block on which the fossiliferous sediments are exposed has been subjected only to rather mild tectonic deformation with some faulting and warping particularly at its lakeward margin (Behrensmeyer, 1970; Vondra *et al.*, 1971; Fitch and Vondra, 1976; Findlater, 1976; Vondra and Bowen, 1976).

Associated finds Stone tools are known from several sites within the research area. A series of core tools, cobbles and flakes identified as belonging to the Oldowan industry have been recovered from the KBS tuff itself and another industry known as the Karari Industry has also been described from the area (Leakey, 1970; Isaac *et al.*, 1971; Isaac, 1976; Isaac *et al.*, 1976; Harris and Isaac, 1976).

Fig. 57   A correlation of collection units, tuffs, potassium-argon dates and faunal zones from the Koobi Fora region
After J. H. Harris, in press

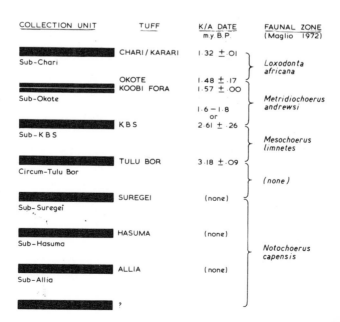

Numerous fossil vertebrates have been recovered from the research area and they have been classified into four faunal assemblages, named after characteristic fossils, to define biostratigraphic zones. These zones have been termed, from older to younger, the *Notochoerus capensis* zone, the *Mesochoerus limnetes* zone, the *Metridiochoerus andrewsi* zone and the *Loxodonta africana* zone. Each zone contains elephants, pigs and hippopotami, and evolutionary trends have been found in these forms through the biostratigraphic sequence. Many other fossil mammals have been recovered including bovids, giraffids, hyaenids, felids and primates (Maglio, 1971 and 1972).

Dating   The dating of the site is based upon stratigraphic, faunal, palaeomagnetic and radiometric data. The sequence of deposits from the oldest to the youngest is nowhere complete but sections of the sequence can be correlated by the tuffs and the fossils contained in the sediments between these layers (see Fig. 56). The radiometric data and the palaeomagnetic data from the site is summarized in Fig. 57 and shows that this interpretation of the stratigraphic column from east of Lake Turkana provides a good fit with the world geomagnetic reversal polarity time scale (Fitch, Findlater, Watkins and Miller, 1974). The upper boundary of the Koobi Fora Formation (Chari-Karari Tuff) is securely dated by K/Ar and fauna to about 1·3 million years B.P. The Kubi Algi Formation appears certain to be about 4–4·5 m.y. old on the basis of its contained fauna and a series of geophysical determinations. The ages of horizon between these limits, especially the KBS tuff is currently the subject of controversy. Fitch and Miller (1970, 1976) using the Ar^{40}/Ar^{39} step heating technique favour an age of 2·6±0·26 for the KBS while Curtis, Drake, Cerling and Hempel (1975) suggest that the 'KBS' is a complex of tuff units with the conventional K/Ar age of one unit being 1·82±0·04 and another 1·60±0·05. Attempts to resolve the uncertainty by other methods such as fission track dating, are in progress. Further details of the dating of the site can be found in Maglio (1972), Brock and Isaac (1974) and Hurford (1974).

The Koobi Fora Remains

Additional results on the palaeomagnetic stratigraphy of the
Koobi Fora Formation have not resolved the dating controversy over the age of the KBS tuff; indeed considerable doubts
have been expressed concerning the published tuff correlations
between areas within the Koobi Fora region (Hillhouse,
Ndombi, Cox and Brock, 1977).

Morphology KNM–ER 406

A virtually complete cranium lacking teeth recovered from
the Ileret Member of the Koobi Fora Formation (Area 10).
The cranium is heavily built with a large face and a small
neurocranium. The facial buttresses are robust with a stout
brow ridge and broad zygomata. The cranial capacity must be
small and there is considerable postorbital constriction. The
cranium possesses a sagittal crest as well as occipital and
supramastoid crests which were probably associated with a
heavy mandible and powerful muscles of mastication (Leakey,
1970; Leakey, Mungai and Walker, 1971; Holloway, 1973).

Fig. 58   The KNM-ER 729 mandible
Occlusal view (cast)

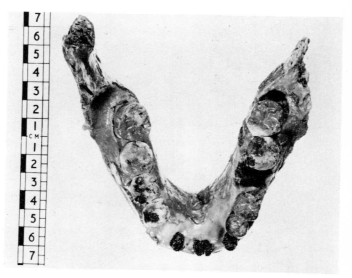

171

KNM–ER 729

A robust mandible with teeth recovered from the Ileret Member of the Koobi Fora Formation (Area 8). The body of the mandible is virtually intact but the ramus on the left is badly eroded and the condyloid process on the right is missing. The body is deep and stout bearing a marked *prominentia lateralis* but no chin. The ramus on the right is tall and buttressed internally by a *crista pharyngea*. The teeth that are present include the central incisors, the canines, the left PM_1, both PM_2 teeth, the right M_1 and the M_2 and M_3 of both sides. The dentition shows marked anteroposterior disproportion and a considerable molarization of the premolars (Leakey, 1971; Leakey, Mungai and Walker, 1972).

KNM–ER 730

Part of a mandibular body recovered from the Upper Member of the Koobi Fora Formation (Area 103), including the right side, the symphysial region and a smaller part of the left side. The crowns of the left molar teeth are retained but heavily worn. The mandible bears a slight chin, a *planum alveolare* and evidence of periodontal disease (Leakey, 1971; Day and Leakey, 1973).

Fig. 59   The KNM–ER 732 A demicranium (cast)

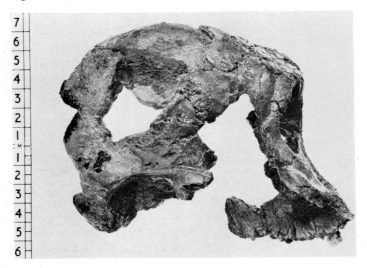

KNM–ER 732A

A gracile demicranium including the right maxilla, much of the frontal and parietals and the right temporal bone recovered from the Ileret Member of the Koobi Fora Formation (Area 10). The vault bones are thin, the maxilla broad and the temporal bone rugged in construction. The supramastoid crest is prominent but the temporal lines are weakly imprinted and do not fuse to form a crest. The only dental remains that are preserved are the roots of the molars and part of the crown of the right P4 (Leakey, 1971; Leakey, Mungai and Walker, 1972; Holloway, 1973).

KNM–ER 736

The shaft of a massive left femur from the Upper Member of the Koobi Fora Formation (Area 103). Superiorly the head, neck and greater trochanter are missing and the lesser trochanter is mostly eroded away; inferiorly the bone is broken across just proximal to the development of the supracondylar lines. The gluteus maximus impression is extensive and includes an hypotrochanteric fossa while the *linea aspera* is well defined. The medial border of the bone is convex medially and the narrowest point on the shaft is low (Leakey, 1971; Leakey, Mungai and Walker, 1972).

KNM–ER 737

The shaft of a left femur from the Upper Member of the Koobi Fora Formation (Area 103). The head and part of the neck are missing, as well as the greater and lesser trochanters, and the bone is broken across distally through the popliteal surface. The bone is characterized by its flatness or platymeria, by the low position of its narrowest point and by the convexity of its medial border (Leakey, 1971; Day and Leakey, 1973).

KNM–ER 739

The shaft and lower end of a large right humerus from the Ileret Member of the Koobi Fora Formation (Area 1). The shaft is marked by a large deltoid impression and a deep groove for the long head of biceps. Distally the lower end is expanded and bears a rounded capitulum and a trochlear surface surmounted by trochlear and capitular fossae. Posteriorly, the olecranon fossa is deep while the medial epicondyle is very pronounced. The lateral border of the shaft distally shows a marked brachioradialis flange (Leakey, 1971; Leakey, Mungai and Walker, 1972).

KNM–ER 803

A partial skeleton recovered from the Ileret Member of the Koobi Fora Formation (Area 8). It includes parts of the shafts of a femur, a tibia, an ulna, fibulae, phalanges, a metatarsal and many other fragments from both left and right sides. The femoral fragment is flattened and tapers distally and shows a pronounced *linea aspera* but no pilaster (Leakey, 1972; Day and Leakey, 1974).

KNM–ER 813

A right talus and part of the shaft of a tibia or a femur from the Upper Member of the Koobi Fora Formation (Area 104). The talus is a little eroded and shows little horizontal angulation of the neck but considerable torsion of the neck (Leakey, 1972; Leakey and Wood, 1973).

KNM–ER 820

A juvenile mandible from the Ileret Member of the Koobi Fora Formation (Area 1) with an almost complete dentition. This pretty little jaw contains elements of both the deciduous and permanent dentitions. The deciduous teeth that are present include the left canine and all four deciduous molars but in addition all four permanent incisors are present as well as both first permanent molars. The incisor teeth are large and spatulate and the first permanent molars show the basic Y5 cusp morphology (Leakey, 1972; Leakey and Wood, 1973).

KNM–ER 992

The two halves of a mandibular body and parts of both rami of a jaw recovered from the Ileret Member of the Koobi Fora Formation (Area 1). The mandible contains all of the molars and premolars as well as parts of both canines and a left incisor. The mandible is broken through the symphysial region. The body of the mandible is relatively gracile and its upper and lower borders are parallel. The tooth row shows a pronounced curve of Spee with no canine projection (Leakey, 1972; Leakey and Wood, 1973).

KNM–ER 999

A massive left femur from the Guomde Formation (Area 6A). The bone retains about half of its head and much of its stout and rounded neck as well as much of the shaft. Muscular markings are pronounced including both a *crista* and a *fossa hypotrochanterica*. The medial condyle of the distal end is present but incomplete, yet retains the semilunar patellar facet. The narrowest point on the shaft appears to be above the mid-point on a minimum length basis (Day and Leakey, 1974).

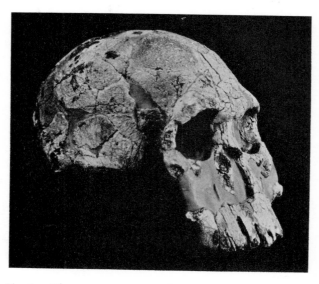

Fig. 60   The KNM-ER 1470 partial cranium.
Reconstructed by Dr A. Walker
Courtesy of the Director of the National Museums of Kenya and Anglia Television Ltd

KNM-ER 1470

A partial cranium from the Lower Member of the Koobi Fora Formation (Area 131). The calvaria has suffered some distortion but with retention of the alignment of midline structures; it is long and ovoid with moderate postorbital constriction. The vault bones are thin but the endocranial surface has suffered some loss of the inner cortical bone. The face is moderately well preserved being broad but with no strong development of the supraorbital ridge and with little or   **175**

The Koobi Fora Remains

no subnasal prognathism. The palate is broad and the alveolar processes stout; erosion of these processes gives a false impression of the shallowness of the palate. The maxillary air sinuses are well developed and project down between the molar roots; the development of these sinuses as well as that of the mastoid air cells suggest that the cranium is that of an adult (Leakey, 1973; Day, Leakey, Walker and Wood, 1974).

KNM-ER 1477

The body of a juvenile mandible with dentition, from the Upper Member of the Koobi Fora Formation (Area 105). The specimen contains part of the deciduous dentition, the first permanent molar germs in open crypts and other tooth germs visible in the mandibular body where its external surface is broken away. The symphysial region shows a distinct post-incisive *planum alveolare* bounded below by a stout superior transverse torus below which are paired genial pits. The first permanent molar germs of both sides are large and possess five main cusps, a Y-fissure pattern and both anterior and posterior foveae (Leakey, 1973; Day, Leakey, Walker and Wood, 1976).

KNM-ER 1481

An associated group of leg bones from the Lower Member of the Koobi Fora Formation (Area 131), including a left femur, proximal and distal left tibial fragments and a distal fibular fragment. The femur is virtually complete and has a large head, a rounded neck, a flared greater trochanter, a prominent lesser trochanter, a well marked *linea aspera* and some degree of shaft flattening. The gluteus maximus impression is present as *fossa hypotrochanterica* with some lateral swelling of the shaft. The narrow point of the shaft appears to be distally placed (Leakey, 1973; Day, Leakey, Walker and Wood, 1974).

KNM-ER 1500

An associated skeleton from the Lower Member of the Koobi Fora Formation (Area 130), including parts of a femur, a tibia, a radius, an ulna, a humerus and a metatarsal as well as many other fragments. The skeleton is that of a small individual whose femur is characterized by marked anteroposterior compression of the neck (Leakey, 1973; Day, Leakey, Walker and Wood, 1976).

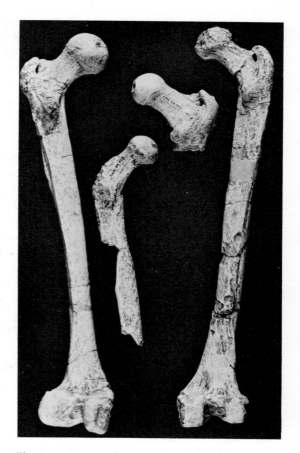

Fig. 61   A group of femora from the Koobi Fora site
From left to right: KNM–ER 1481, KNM–ER 738,
KNM–ER 1503 and KNM–ER 1475 Posterior views
*Courtesy of the Director of the National Museums
of Kenya and Dr B. A. Wood*

KNM–ER 1503

An almost perfect proximal end of a right femur from the
Koobi Fora Formation (Area 123). The head of the bone is
small and the neck long and flattened anteroposteriorly. There
is no *linea intertrochanterica* marking the attachment of the
vertical fibres of the iliofemoral ligament, but there is a   177

prominent femoral tubercle. The posterior aspect of the neck is marked by a broad shallow groove for the obturator externus tendon (Leakey, 1973; Day, Leakey, Walker and Wood, 1976).

KNM–ER 1590

A partial cranium with a juvenile dentition recovered from the Lower Member of the Koobi Fora Formation just below the KBS tuff (Area 12). It consists of both parietals and fragments of frontal of an immature skull associated with deciduous and permanent teeth. The cranial vault is thin but suggests that the cranial volume was large. The teeth are well preserved and include a permanent left first incisor, the left deciduous canine and second deciduous molar, both unerupted canines and premolars, the erupted left and right first molars and the left second molar (Leakey, 1974; Day, Leakey, Walker and Wood, 1976).

KNM–ER 1802

A beautifully preserved body of an adult mandible from the Lower Member of the Koobi Fora Formation (Area 133). The body is preserved from the distal end of the second molar on the left through the symphysis to the inter-alveolar septum between the second and third molars on the right. The symphysis is thick, but there are no superior or inferior transverse tori. The first molar teeth of both sides are present and show six cusps arranged in a Y pattern; the second molars are also present and are larger than the first molars. No significant curve of Spee has developed (Leakey, 1974; Day, Leakey, Walker and Wood, 1976).

KNM–ER 1805

A cranium and mandible recovered from the Upper Member of the Koobi Fora Formation (Area 130). The cranium is almost complete, heavily built and possesses a distinct sagittal crest and a marked supramastoid crest on the left. The face is represented by the maxillae and nasal bones and the palate is almost complete although somewhat splayed and contains an almost complete dentition. The body of the mandible is moderately robust and contains the right second and third molar teeth and the roots of many others. The rami are missing (Leakey, 1974; Day, Leakey, Walker and Wood, 1976).

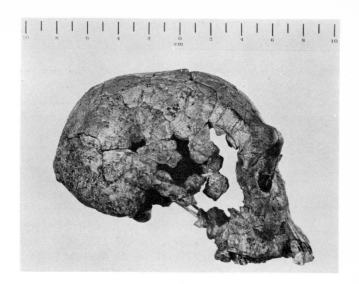

Fig. 62   The KNM-ER 1813 cranium
Right lateral view
Courtesy of the Director of the National Museums of Kenya

Fig. 63   The KNM-ER 1813 palate and
dentition
*Courtesy of the Director of the
National Museums of Kenya*

KNM-ER 1813

A cranium recovered from the Koobi Fora Formation (Area
123). The cranium is virtually complete, missing only part of
the base, the left zygomatic region and part of the left orbit. 179

The Koobi Fora Remains

The dentition includes the left canine, both left premolars and all three molars on the left as well as the right incisors, canine and second and third molars. The skull is small, lightly built and with no cranial cresting. The supraorbital ridges are modest and the face shows a small degree of nasal prominence (Leakey, 1974; Day, Leakey, Walker and Wood, 1976).

KNM-ER 3228

A right hip bone (*os coxa*) from the Lower Member of the Koobi Fora Formation (Area 102). The bone is almost complete missing only the pubic portion. The bone is robustly constructed with a broad iliac flange, an acute sciatic notch, an hyper-robust iliac pillar and a laterally rotated ischium. The acetabular fossa is large, indicating a large femoral head (Leakey, 1976).

KNM-ER 3733

A remarkably preserved cranium from the Upper Member of the Koobi Fora Formation (Area 104). The vault is complete and undistorted and much of the facial skeleton is present as well as examples of the premolar and molar teeth. The cranial capacity seems likely to be of the order of 800–900 cc. The preliminary description of the specimen (Leakey, 1976) makes it clear that there is little doubt that this skull must be attributed to *Homo erectus*. Its similarities to the Peking skulls are striking. It seems likely to be the best preserved example of the skull of *Homo erectus* from anywhere in the world (Leakey, 1976).

Dimensions   The dimensions of the East Rudolf hominid material described in this section are to be found in the descriptive publications referred to in relation to each specimen.

Affinities   At the present time all of the material from this site has been placed in the family Hominidae; of this some has been placed in the genus *Australopithecus*, some into the genus *Homo* and the remainder left *incertae sedis* (see table below). As a matter of policy, the hominid team of the East Rudolf Research Project has usually refrained from making considered specific allocations or erecting new taxa until the material has been published in detail to avoid the confusions that can arise from premature systematic assertions, while ensuring the 180 publication of new finds as rapidly as possible. Naturally this policy relies heavily on the co-operation of scholars the world over, co-operation that has for the most part been given readily.

KOOBI FORA HOMINIDS

KNM-ER No.	Year	Area	Specimen	Stratigraphic Position	Published Taxonomic Status
164A, B & C	1969/71	104	Parietal frag., 2 phalanges 2 vertebrae	Upper Memb. K.F. formation above KBS tuff	Homo
403	1968	103	Rt. ½ mandibular body	Upper Memb. K.F. formation below Koobi Fora tuff	Australopithecus
404	1968	7A	Rt. ½ mandibular body	Ileret Memb. K.F. formation below Chari tuff	Australopithecus
405	1968	105	Palate lacking teeth	Upper Memb. K.F. formation above post-KBS erosional surface	Australopithecus
406	1969	10	Cranium lacking teeth	Ileret Memb. K.F. formation below Okote tuff complex	Australopithecus
407	1969	10	Partial calvaria lacking face	Ileret Memb. K.F. formation below Okote tuff complex	cf. Australopithecus
417	1968	129	Parietal frag.	Upper Memb. K.F. formation	Australopithecus
725	1970	1	Lt. ½ mandibular body	Ileret Memb. K.F. formation just above Okote tuff complex	Australopithecus
726	1970	10	Lt. ½ mandibular body	Ileret Memb. K.F. formation just above Okote tuff complex	Australopithecus
727	1970	6A	Frag. rt. ½ mandibular body	Ileret Memb. K.F. formation just below Okote tuff complex	Australopithecus
728	1970	1	Rt. ½ mandibular body	Ileret Memb. K.F. formation just above Okote tuff complex	Australopithecus
729	1970	8	Mandible with dentition	Ileret Memb. K.F. formation lower part of Okote tuff complex	Australopithecus
730	1970	103	Mandible lacking rt. side post. to P₄, lt. M₁–M₃	Upper Memb. K.F. formation just below Koobi Fora tuff	Homo
731	1970	6A	Lt. ½ mandibular body lacking teeth	Ileret Memb. K.F. formation below Chari tuff	Homo
732A	1970	10	Demi-cranium	Ileret Memb. K.F. formation below Okote tuff complex	Australopithecus
733	1970	8	Rt. ½ mandibular body, lt. maxilla, cranial frags.	Ileret Memb. K.F. formation in Okote tuff complex	Australopithecus
734	1970	103	Parietal frag.	Upper Memb. K.F. formation just below Koobi Fora tuff	Incertae sedis
736	1970	103	Lt. femoral shaft	Upper Memb. K.F. formation below projected level of Koobi Fora tuff	Australopithecus or early Homo
737	1970	103	Lt. femoral shaft	Upper Memb. K.F. formation in base of Koobi Fora tuff	Homo
738	1970	105	Prox. end lt. femur	Upper Memb. K.F. formation above post KBS erosional surface	Australopithecus
739	1970	1	Rt. humerus	Ileret Memb. K.F. formation just above Okote tuff complex	Australopithecus
740	1970	1	Frag. distal end lt. humerus	Ileret Memb. K.F. formation	Australopithecus
741	1970	1	Prox. end rt. tibia	Ileret Memb. K.F. formation above Okote tuff complex	Australopithecus or early Homo
801	1971	6A	Rt. ½ mandibular body and assoc. isolated teeth	Ileret Memb. K.F. formation below Okote tuff complex	Australopithecus
802	1971	6A	Isolated teeth	Ileret Memb. K.F. formation below Okote tuff complex	Australopithecus
803	1971	8	Associated skeletal elements	Ileret Memb. K.F. formation in Okote tuff complex	Homo

KNM-ER No.	Year	Area	Specimen	Stratigraphic Position	Published Taxonomic Status
805	1971	I	Frag. lt. ½ mandibular body	Ileret Memb. K.F. formation above Okote tuff complex	*Australopithecus*
806	1971	8	Isolated teeth	Ileret Memb. K.F. formation in Okote tuff complex	*Homo*
807A	1971/73	8A	Frag. rt. maxilla, M³ part M² Frag. rt. maxilla, M¹	Ileret Memb. K.F. formation in Okote tuff complex	*Homo*
808	1971	8	Isolated juvenile teeth	Ileret Memb. K.F. formation in Okote tuff complex	*Homo*
809	1971	8	Isolated teeth	Ileret Memb. K.F. formation in Okote tuff complex	*Homo*
810	1971	104	Lt. ½ mandibular body, small part of rt. side, M₃	Upper Memb. K.F. formation above KBS tuff	*Australopithecus*
811	1971	104	Parietal frag.	K.F. formation, probably below KBS tuff	*Homo*
812	1971	104	Frag. lt. ½ juvenile mandible	Upper Memb. K.F. formation above KBS tuff	*Australopithecus*
813	1971	104	Rt. talus and frag. distal end rt. tibia/lt. femur	Upper Memb. K.F. formation above KBS tuff	*Homo*
814	1971	104	Cranial frags.	Upper Memb. K.F. formation above KBS tuff	*Australopithecus*
815	1971	10	Frag. prox. end lt. femur	Ileret Memb. K.F. formation below Okote tuff complex	*Australopithecus*
816	1971	104	Lt. upper canine and molar frags.	Upper Memb. K.F. formation above KBS tuff	*Australopithecus*
817	1971	124	Frags. lt. ½ mandibular body	Koobi Fora formation, exact position unclear	*Homo*
818	1971	6A	Lt. ½ mandibular body, P₃–M₃	Ileret Memb. K.F. formation within Okote tuff complex	*Australopithecus*
819	1971	I	Lt. ½ mandibular body	Ileret Memb. K.F. formation just below Okote tuff complex	*Australopithecus*
820	1971	I	Juvenile mandible with dentition	Ileret Memb. K.F. formation in lower part of Okote tuff complex	*Homo*
992	1971	I	Mandible with dentition	Ileret Memb. K.F. formation above Okote tuff complex	*Homo*
993	1971	I	Distal ¾ rt. femur	Ileret Memb. K.F. formation just above Okote tuff complex	*Australopithecus*
997	1971	104	Prox. end lt. metatarsal III	Upper Memb. K.F. formation above KBS tuff	*Australopithecus*
998	1971	104	Rt. I²	Upper Memb. K.F. formation above KBS tuff	*Australopithecus*
999	1971	6A	Lt. femur, frag. femoral condyle, isolated frags.	Guonde Formation	*Homo*
1170	1971	6A	Cranial frags.	Ileret Memb. K.F. formation below Okote tuff complex	*Australopithecus*
1171	1971	6A	Isolated juvenile teeth	Ileret Memb. K.F. formation below Okote tuff complex	*Australopithecus*
1462	1972	130	Isolated lt. M₃	Lower Memb. K.F. formation below KBS tuff	*Homo*
1463	1972	I	Rt. femoral shaft	Ileret Memb. K.F. formation in Okote tuff complex	*Australopithecus*
1464	1972	6A	Rt. talus	Ileret Memb. K.F. formation below Okote tuff complex	*Australopithecus* or early *Homo*
1465	1972	II	Frag. prox. end lt. femur	Ileret Memb. K.F. formation below Chari tuff	*Australopithecus*
1466	1972	6	Fronto-parietal frag.	Ileret Memb. K.F. formation above Okote tuff complex	*Homo*
1467	1972	3	Isolated lt. M₃	Ileret Memb. K.F. formation above Okote tuff complex	*Australopithecus*

KNM-ER No.	Year	Area	Specimen	Stratigraphic Position	Published Taxonomic Status
1468	1972	11	Rt. ½ mandibular body	Ileret Memb. K.F. formation just in or above Okote tuff complex	Australopithecus
1469	1972	131	Lt. ½ mandibular body, M_2–M_3	Lower Memb. K.F. formation below KBS tuff	Australopithecus
1470	1972	131	Cranium	Lower Memb. K.F. formation below KBS tuff	Homo
1471	1972	131	Prox. ⅓ rt. tibia	Lower Memb. K.F. formation below KBS tuff	Australopithecus or early Homo
1472	1972	131	Rt. femur	Lower Memb. K.F. formation below KBS tuff	Homo
1473	1972	131	Prox. end rt. humerus	Lower Memb. K.F. formation below KBS tuff	Incertae sedis
1474	1972	131	Parietal frag.	Lower Memb. K.F. formation below KBS tuff	Incertae sedis
1475	1972	31	Prox. end rt. femur	Lower Memb. K.F. formation below KBS tuff	Homo
1476	1972	105	Lt. talus, prox. end lt. tibia, frag. shaft rt. tibia	Upper Memb. K.F. formation above post KBS erosional surface	Australopithecus
1477	1972	105	Juv. mandible with dentition	Upper Memb. K.F. formation above post KBS erosional surface	Australopithecus
1478	1972	105	Cranial frag.	Upper Memb. K.F. formation above post KBS erosional surface	Australopithecus
1479	1972	105	Isolated tooth frags.	Upper Memb. K.F. formation above post KBS erosional surface	Australopithecus
1480	1972	105	Isolated rt. M_3	Upper Memb. K.F. formation above post KBS erosional surface	Homo
1481	1972	131	Lt. femur, prox and dist. ends lt. tibia, dist. end lt. fibula	Lower Memb. K.F. formation below KBS tuff	Homo
1482	1972	131	Mandible, and isolated teeth	Lower Memb. K.F. formation below KBS tuff	Incertae sedis
1483	1972	131	Mandibular frags.	Lower Memb. K.F. formation below KBS tuff	Homo
1500	1972	130	Associated skeletal elements	Lower Memb. K.F. formation just below KBS tuff	Australopithecus
1501	1972	123	Rt. ½ mandibular body	Upper Memb. (provisional) K.F. formation	Homo
1502	1972	123	Frag. rt. ½ mandibular body, M_1	Upper Memb. (provisional) K.F. formation	Homo
1503	1972	123	Prox. end rt. femur	Upper Memb. (provisional) K.F. formation	Australopithecus
1504	1972	123	Frag. dist. end rt. humerus	Upper Memb. (provisional) K.F. formation	Australopithecus
1505	1972	123	Prox. end lt. femur, frag. dist. end lt. femoral shaft	Upper Memb. (provisional) K.F. formation	Australopithecus
1506	1972	121	Frag. rt. ½ mandibular body, M_1, M_2, isolated P^3, P^4	Upper Memb. (provisional) K.F. formation	Australopithecus
1507	1972	127	Lt. ½ juvenile mandibular body	Upper Memb. K.F. formation	Homo
1508	1972	127	Isolated rt. molar	Upper Memb. K.F. formation	Homo
1509	1972	119	Isolated teeth C–M_3	Upper Memb. K.F. formation	Australopithecus
1515	1972	103	Isolated rt. I^2	Upper Memb. K.F. formation below Koobi Fora tuff	Incertae sedis
1590	1972	12	Partial juvenile cranium with dentition	Lower Memb. K.F. formation below KBS tuff	Homo
1591	1972	12	Rt. humerus lacking head	Ileret Memb. K.F. formation above KBS tuff	Homo

KNM-ER No.	Year	Area	Specimen	Stratigraphic Position	Published Taxonomic Status
1592	1972	12	Dist. ½ rt. femur	Ileret Memb. K.F. formation above KBS tuff	*Australopithecus*
1593	1972	12	Parietal and mandibular frags.	At base of Ileret Memb. K.F. formation at projected level of KBS tuff	*Homo*
1648	1971	105	Parietal frag.	Upper Memb. K.F. formation	*Incertae sedis*
1800	1973	130	Cranial frags.	Lower Memb. K.F. formation below KBS tuff	*Incertae sedis*
1801	1973	131	Lt. ½ mandibular body, P_4, M_1, M_3	Lower Memb. K.F. formation below KBS tuff	cf. *Homo*
1802	1973	131	Mandible, tooth frags.	Lower Memb. K.F. formation below KBS tuff	cf. *Homo*
1803	1973	131	Frag. rt. ½ mandibular body	Lower Memb. K.F. formation below KBS tuff	*Incertae sedis*
1804	1973	104	Frag. rt. maxilla, P^3–M^2	Upper Memb. K.F. formation above KBS tuff	*Incertae sedis*
1805	1973	130	Cranium and mandible with dentition	Upper Memb. K.F. formation below Okote tuff complex	*Incertae sedis*
1806	1973	130	Mandible, no teeth	Upper Memb. K.F. formation below Okote tuff complex	cf. *Australopithecus*
1807	1973	103	Dist. ⅔ femoral shaft	Upper Memb. K.F. formation above Koobi Fora tuff	*Incertae sedis*
1808	1973	103	Associated skeletal and cranial elements	Upper Memb. K.F. formation below Koobi Fora tuff	*Incertae sedis*
1809	1973	121	Rt. femoral shaft	Upper Memb. K.F. formation (provisional)	*Incertae sedis*
1810	1973	123	Prox. end lt. tibia	Upper Memb. (provisional) K.F. formation	*Incertae sedis*
1811	1973	123	Mandibular frags.	Upper Memb. (provisional) K.F. formation	*Incertae sedis*
1812	1973	123	Frag. rt. ½ mandibular body isolated lt. I_2, M, head of radius	Upper Memb. (provisional) K.F. formation	*Incertae sedis*
1813	1973	123	Cranium, dentition	Upper Memb. (provisional) K.F. formation	cf. *Australopithecus*
1814	1973	127	Associated elements of a lower dentition	Upper Memb. (provisional) K.F. formation	*Incertae sedis*
1816	1973	6A	Frags. juv. mandible	Ileret Memb. K.F. formation below Okote tuff complex	*Incertae sedis*
1817	1973	1	Lt. ½ mandibular body	Ileret Memb. K.F. formation below Okote tuff complex	*Incertae sedis*
1818	1973	6A	Isolated rt. I^1	Ileret Memb. K.F. formation below Okote tuff complex	*Incertae sedis*
1819	1973	3	Isolated lt. M_3 crown	Ileret Memb. K.F. formation below Okote tuff complex	*Incertae sedis*
1820	1973	103	Frag. lt. ½ mandibular body dm_2 and M_1	Upper Memb. K.F. formation below Koobi Fora tuff	*Incertae sedis*
1821	1973	123	Parietal frag.	Upper Memb. (provisional) K.F. formation	*Incertae sedis*
1822	1973	123	Frag. femoral shaft	Upper Memb. (provisional) K.F. formation	*Incertae sedis*
1823	1971	6A	Prox. end metatarsal	Ileret Memb. K.F. formation below Okote tuff complex	*Incertae sedis*
1824	1971	6A	Dist. end rt. humerus frag.	Ileret Memb. K.F. formation below Okote tuff complex	*Incertae sedis*
1825	1971	6A	Frag. atlas	Ileret Memb. K.F. formation below Okote tuff complex	*Incertae sedis*
2592	1974	6	Parietal frag.	Ileret Memb. K.F. formation probably below Okote tuff complex	In press

KNM-ER No.	Year	Area	Specimen	Stratigraphic Position	Published Taxonomic Status
2593	1974	6	Molar frag.	Ileret Memb. K.F. formation in or above Okote tuff complex	In press
2595	1974	1A	Parietal frag.	Ileret Memb. K.F. formation probably below Okote tuff complex	In press
2596	1974	15	Dist. end lt. tibia	Upper Memb. K.F. formation above KBS tuff	In press
2597	1974	15	Lower lt. molar (M_2 or M_3)	K.F. formation approx. level KBS tuff	In press
2598	1974	15	Occipital frag.	K.F. formation approx. level KBS tuff	In press
2599	1974	15	Frag. lt. P_4	K.F. formation approx. level KBS tuff	In press
2600	1974	130	Half molar	Lower Memb. K.F. formation below KBS tuff	In press
2601	1974	130	Crown, rt. lower molar	Lower Memb. K.F. formation below KBS tuff	In press
2602	1974	117	Cranial frags.	Lower Memb. K.F. formation just above Tulu Bor tuff	In press
2603	1974	117	Tooth frag.	Lower Memb. K.F. formation below Tulu Bor tuff	In press
2604	1974	117	Tooth	Lower Memb. K.F. formation just above Tulu Bor tuff	In press
2605	1974	117	Tooth frag.	Lower Memb. K.F. formation below Tulu Bor tuff	In press
2606	1974	117	Tooth frag.	Lower Memb. K.F. formation below Tulu Bor tuff	In press
2607	1972	105	Molar frag.	Upper Memb. K.F. formation above KBS tuff	Homo (first published as KNM-ER 1480B)
3228	1975	102	Rt. innominate	Lower Memb.	In press
3229	1975	103	Mandible, rt. P_3, lt. P_4	Upper Memb.	In press
3230	1974	130	Mandible with dentition	Upper Memb. K.F. formation in Okote tuff complex	In press
3728	1975	100	Shaft and neck of right femur	Lower Memb. below KBS tuff	—
3729	1975	102	Body of left mandible, partial roots C–M_3	? Upper Memb.	—
3730	1975	102	Weathered distal femur	? Upper Memb.	—
3731	1975	105	Body of left mandible, partial roots I–M_3	Lower Member, below KBS tuff	—
3732	1975	105	Partial cranium	Lower Memb. below KBS tuff	—
3733	1975	104	Cranium with partial dentition	Upper Member, below Koobi Fora tuff	Homo erectus
3734	1975	105	Body of left mandible, C–M_2, frag. of M_3	Lower Member, below KBS tuff	—
3735	1975	116	Weathered distal right humerus	Lower Memb. below KBS tuff	—
3736	1975	105	Proximal ⅔ of a radius	Upper Memb. below KBS tuff	—
3737	1971	6A	Frags. rt. M_3 and M_1	Ileret Memb. K.F. formation below Okote tuff complex	Australopithecus (first published as KNM-ER 802)

The Koobi Fora Remains

Originals The National Museum of Kenya, Nairobi, Kenya.

Casts The National Museum of Kenya, Nairobi, Kenya.

References BEHRENSMEYER, A. K. 1970
Preliminary Geological Interpretation of a New Hominid Site in the Lake Rudolf Basin. *Nature 226,* 225-226.
FITCH, F. J. and MILLER, J. A. 1970
Radioisotopic Age Discriminations of Lake Rudolf Artefact Site. *Nature 226,* 226-228.
LEAKEY, R. E. F. 1970
New Hominid Remains and Early Artefacts from Northern Kenya. *Nature 226,* 223-224.
ISAAC, G. L., LEAKEY, R. E. F. and BEHRENSMEYER, A. K. 1971
Archaeological traces of early hominid activities, east of Lake Rudolf, Kenya. *Science N.Y. 173,* 245-248.
LEAKEY, R. E. F. 1971
Further Evidence of Lower Pleistocene Hominids from East Rudolf, North Kenya. *Nature 231,* 241-245.
LEAKEY, R. E. F., MUNGAI, J. M. and WALKER, A. C. 1971
New Australopithecines from East Rudolf, Kenya. *Am. J. Phys. Anthrop. 35,* 175-186.
MAGLIO, V. J. 1971
Vertebrate Faunas from the Kubi Algi, Koobi Fora and Ileret Areas, East Rudolf, Kenya. *Nature 231,* 248-249.
VONDRA, C. F., JOHNSON, G. D., BOWEN, B. E. and BEHRENSMEYER, A. K. 1971
Preliminary Stratigraphical Studies of the East Rudolf Basin, Kenya. *Nature 231,* 245-248.
LEAKEY, R. E. F. 1972
Further Evidence of Lower Pleistocene Hominids from East Rudolf, North Kenya 1971. *Nature 237,* 264-269.
LEAKEY, R. E. F., MUNGAI, J. M., and WALKER, A. C. 1972
New Australopithecines from East Rudolf Kenya (11). *Am. J. Phys. Anthrop. 36,* 235-251.
MAGLIO, V. J. 1972
Vertebrate Faunas and Chronology of Hominid-bearing Sediments East of Lake Rudolf, Kenya. *Nature 239,* 379-385.
LEAKEY, R. E. F. 1973
Further Evidence of Lower Pleistocene Hominids from East Rudolf, North Kenya 1972. *Nature 242,* 170-173.
DAY, M. H. and LEAKEY, R. E. F. 1973
New evidence for the genus *Homo* from East Rudolf, Kenya. 1. *Am. J. Phys. Anthrop. 39,* 341-354.
LEAKEY, R. E. F., and WOOD, B. A. 1973
New evidence for the genus *Homo* from East Rudolf, Kenya 11. *Am. J. Phys. Anthrop. 39,* 355-368.

HOLLOWAY, R. L. 1973
New endocranial values for the East African early hominids. *Nature*
243, 97–99.
BROCK, A. and ISAAC, G. L. 1974
Palaeomagnetic stratigraphy and chronology of hominid-bearing
sediments east of Lake Rudolf, Kenya. *Nature, 247,* 344-348.
DAY, M. H., and LEAKEY, R. E. F. 1974
New evidence for the genus *Homo* from East Rudolf, Kenya III.
Am. Phys. Anthrop. 41, 367–380.
DAY, M. H., LEAKEY, R. E. F., WALKER, A. C. and WOOD, B. A. 1974
New Hominids from East Rudolf, Kenya, I. *Am. J. Phys. Anthrop.*
42, 461-476.
FITCH, F. J., FINDLATER, I. C., WATKINS, R. T. and MILLER, J. A. 1974
Dating of the Rock Succession containing fossil hominids at East
Rudolf, Kenya. *Nature 251*, 213-215.
HURFORD, A. J. 1974
Fission track dating of a vitric tuff from East Rudolf, Kenya. *Nature*
249, 236-237.
LEAKEY, R. E. F. 1974
Further Evidence of Lower Pleistocene hominids from East Rudolf,
North Kenya, 1973. *Nature 248*, 653-656.
CURTIS, G. H., DRAKE, T. CERLING and HAMPEL 1975 [*sic*]
Age of KBS tuff in Koobi Fora Formation, East Rudolf, Kenya.
Nature 258, 395-398.
GROVES, C. P. and MAZÁK, V. 1975
An approach to the taxonomy of the Hominidae: Gracile Villa-
franchian Hominids of Africa. *Cas. Miner. Geol. 20*, 225-247.
DAY, M. H., LEAKEY, R. E. F., WALKER, A. C. and WOOD, B. A. 1976
New hominids from East Turkana, Kenya. *Am. J. Phys. Anthrop.*,
45, 369-436.

FINDLATER, I. C. 1976
Tuffs and the recognition of isochronous mapping units in the Rudolf
succession. In *Earliest man and environments in the Lake Rudolf basin:
stratigraphy, paleoecology and evolution.* Eds. Y. Coppens, F. C.
Howell, G. Ll. Isaac and R. E. F. Leakey. Chicago: Chicago Univer-
sity Press.
FITCH, F. J. and MILLER, J. A. 1976
Conventional Potassium-Argon and Argon-40/Argon-39 dating of
the volcanic rocks from East Rudolf. *Ibid.*
FITCH, F. J. and VONDRA, C. F. 1976
'Tectonic Framework'. *Ibid.*
HARRIS, J. W. K. and ISAAC, G. Ll. 1976
The Karari industry: early Pleistocene archaeological evidence from
the terrain east of Lake Turkana, Kenya. *Nature 262*, 102–107.
ISAAC, G. L. 1976
Plio-Pleistocene artefact assemblages from East Rudolf, Kenya. In
Earliest man and environments in the Lake Rudolf basin: stratigraphy,

The Koobi Fora Remains

paleoecology and evolution. Eds. Y. Coppens, F. C. Howell, G. Ll.
Isaac and R. E. F. Leakey. Chicago: Chicago University Press.
ISAAC, G. L., HARRIS, J. W. K. and CRADER, D. 1976
Archaeological evidence from the Koobi Fora formation. *Ibid.*
LEAKEY, R. E. F. 1976
New hominid fossils from the Koobi Fora formation in Northern
Kenya. *Nature 261,* 574-576.
VONDRA, C. F. and BOWEN, B. E. 1976
Plio-Pleistocene deposits and environments, East Rudolf, Kenya. In
*Earliest man and environments in the Lake Rudolf basin: stratigraphy,
paleoecology and evolution.* Eds. Y. Coppens, F. C. Howell, G. L. Isaac
and R. E. F. Leakey. Chicago: Chicago University Press.
HILLHOUSE, J. W., NDOMBI, J. W. M., COX, A. and BROCK, A. 1977
Additional results on palaeomagnetic stratigraphy of the Koobi Fora
formation, east of Lake Turkana (Lake Rudolf), Kenya. *Nature 265,*
411-415.

The Omo Remains

Synonyms and other names	(1) *Paraustralopithecus aethiopicus* (Arambourg and Coppens, 1967) (2) *Australopithecus cf. africanus* (Howell, 1969) *Australopithecus cf. boisei* (Howell, 1969) *Australopithecus sp.* (Howell, 1969) *Australopithecus boisei* (Howell and Wood, 1974) (3) *Homo sapiens* (Day, 1969)
Site	The lower basin of the Omo river, south-west Ethiopia.
Found by	The International Palaeontological Research Expedition to the Omo Valley. (1) Material recovered by the French team. (2) Material recovered by the American team. (3) Material recovered by the Kenyan team, 1967.
Geology	Five sedimentary formations have been formally recognized within the lower Omo basin, a tectonic depression that forms the northern part of the Lake Turkana trough. The Mursi, Nkalabong, Usno and Shungura Formations comprise the Omo group and are primarily fluvial deposits consisting of cross-bedded sands, silts and clays, interspersed with several pebble gravel lenses. Throughout each formation the fluvial deposits grade both laterally and stratigraphically into sediments of deltaic and lacustrine origin. The depositional environment of these deposits was probably not unlike that of the present day in relation to the Omo River, its delta and Lake Turkana into which it extends. Interbedded within these sediments are tuffaceous beds. In the Shungura Formation, the most extensive formation within the basin, thirteen principal tuffs (designated A–M upwards) have been recognized. In addition to containing elements capable of being dated by the potassium-argon technique, these tuffs provide a useful series of marker beds for further dividing the sequence into members which are named according to their under-lying tuff. The last of the five formations is the much younger Kibish Formation which unconformably overlies the older formations in all regions of the basin. It is upon this formation that the present-day land surface is formed. The French and

190

American teams recovered hominids from the Usno Formation at Brown Sands and White Sands as well as from Members B, C, D, E, F, G, H and J of the Shungura Formation. The Kenyan team worked in more recent deposits in the Kibish Formation which is divided into four Members (I–IV); hominid fossils were recovered from the oldest of these Members, Member I.

Associated finds Both the American and French teams recovered stone artefacts from their respective areas, many made of quartz and all small flake fragments, small pebble fragments and angular fragments some of which show signs of utilization and retouch. No large tools were found that compare with those from Olduvai Gorge (*q.v.*).

The Kenyan team recovered very few tools from the Kibish hominid sites, all of which were surface finds except for a few flakes that were found *in situ*. (Merrick, de Heinzelin, Haesaerts and Howell, 1973; Coppens, Chavaillon and Beden, 1973).

Large numbers of fossil mammals have been recovered from the Omo region including elephants, pigs, bovids, cercopithecoid primates, hippopotami, rhinoceroses and carnivores. (Arambourg, Chavaillon and Coppens, 1967; Howell and Coppens, 1973; Howell, Fichter and Eck, 1969).

Fossil mammals recovered from the Kibish Formation include buffalo, rhinoceros, elephant and cercopithecoid primates.

Dating The lower Omo valley contains a long sequence of sedimentary and pyroclastic deposits of Plio/Pleistocene age (Brown, 1971) as well as deposits of later Pleistocene and Holocene age (Butzer, 1971). The Omo deposits have yielded a fossil vertebrate assemblage from a long sequence of deposits which affords many opportunities for both radiometric dating and palaeomagnetic correlation.

The relatively unfossiliferous Mursi and Nkalabong Formations span the early half of the Pliocene epoch. The basaltic lava flow at the top of the Mursi Formation has a potassium-argon age of c. 4·05 m.y. B.P., and within Member II of the Nkalabong Formation are a series of tuffs which have been

radiometrically dated at c. 3·95 m.y. B.P. The basalt Member
II of the Usno Formation has been dated at 3·31 m.y. B.P., and
the lower tuff of the Triple Tuff sequence, Member VII, is
dated at 2·64 m.y. B.P., with the most fossiliferous horizon
being about 2·9 m.y. B.P. in age. Both the faunal and radiometric dating indicates that the Usno Formation spans the
later half of the Pliocene and is contemporary with the deposits
in the Shungura Formation below Member D.

The Shungura Formation has produced a substantial number
of vertebrate fossils, most of them coming from fluviatile
Members C to the lower Member G. The revised potassium-
argon dating (Brown, in press) places the age for Tuff B in
the lower part of the section at 3·2 m.y. B.P. and Tuff L (the
youngest dated tuff) at 1·3 m.y. B.P. The palaeomagnetic
evidence from this formation is in agreement with the
potassium-argon age estimates for the various tuffs and
inclusive members.

Mursi Formation	4·05–4·4 m.y. B.P.
Nkalabong Formation	3·95 m.y. B.P.
Shungura Formation	1·82–3·75 m.y. B.P.
Kibish Formation	3100–130,000 yr. B.P.

(after Butzer, 1971)

More recent data for the Omo group of Formations have
been given as Mursi/Nkalabong 4·4–<3·9 (Butzer, 1974)
and Shungura >3·7–1·2 (de Heinzelin and Haesaerts, 1974).

Morphology (1) A mandibular body containing tooth roots that indicate
small incisors and canines but large molars and premolars. The
rami of the mandible are missing and the body is extremely
robust. The *planum alveolare* is small and the genioglossal
fossae are deep; the symphysis is very thick as are the right
and left sides of the body in the region of the molar teeth.

(2) SH–7A–125

The body of a robust mandible containing all of the teeth
except three incisors. The mandibular body is massive with no
chin and a deep and stout symphysis. The dentition displays
marked anteroposterior disproportion with heavy molarization of the premolars and reduction of the canines and
incisors.

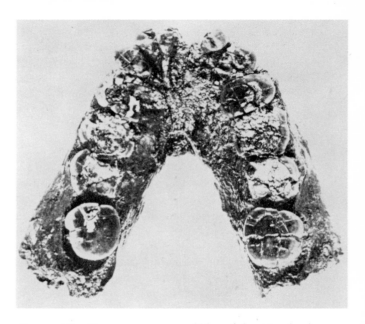

Fig. 64   The Omo SH-7A-125 mandible and dentition (cast)
Occlusal view
Courtesy of F. Clark Howell

SH-74A-21

The right side of a mandibular body containing the canine
and the P_4 as well as the roots of the P_3 and the first molar.
The body of the mandible is comparatively slender but deep;
the ramus is missing other than a small part of its root. The
fourth premolar shows some degree of molarization but heavy
attrition has removed much of the cusp detail.

SH-40-19

An almost complete right ulna that is long and attenuated
with shaft curvature that is marked and convex posteriorly.
Features that depart from the human condition include its
length and curvature, its cross-sectional shaft profile, and
some features of the head of the bone and of the muscle
attachment pattern. Overall these features seem to point to
an elongated forearm with some adaptions not unlike those   193

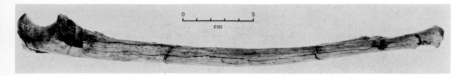

Fig. 65 The Omo SH-40-19 right ulna
Lateral view
Courtesy of B. A. Wood

seen in modern knuckle-walkers; however, features that are unlike those of knuckle-walkers are the 'set' of the articular surface of the upper end and the lack of buttressing of the coronoid process.

(3) OMO I

Site KHS produced a partial skeleton that was found partly *in situ* from the level of a minor unconformity within Member I of the Kibish Formation. It consists of an incomplete vault, parts of the mandible and both maxillae and two tooth crowns (a right upper canine and a left lower first molar). The postcranial remains are extensive and include parts of the skeleton of the upper limb girdle, the arm, forearm and hand as well as parts of the vertebral column. The lower limb remains include parts of the right femur, both tibiae, the right fibula and the right foot.

The skull is robust in construction with a rounded vault, an expanded parietal region and a restricted nuchal plane. The mandibular and maxillary fragments show a rounded dental arcade while the symphysial portion of the mandible has a well marked chin. The teeth are robust but worn. The postcranial skeleton is essentially sapient in its general morphology.

OMO II

This calvaria was recovered from Site PHS as a surface find and it is an almost intact cranial vault with much of the base retained. The skull is heavily built with stout parietes. The form of the vault is dolichocephalic with a receding forehead, a striking occipital torus and an extensive flattened nuchal plane. In frontal view the low vault is marked by a sessile keel with parasagittal flattenings and the maximum breadth of the 194

The Omo Remains

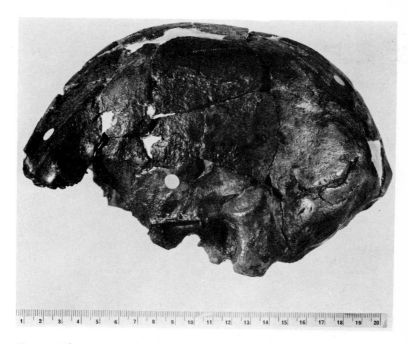

Fig. 66   The Omo II calvaria
Left lateral view

vault is low on the mastoid portion of the temporal bone. The mastoid processes are large and downturned while the articular fossae are deep.

OMO III

This specimen consists of a glabellar fragment and a fronto-parietal vault fragment. The glabellar fragment is heavily built with some evidence of a broad brow and frontal recession.

Dimensions   (1) MANDIBLE
Arambourg and Coppens (1968)
Total length of fragment   75
Height of body at M_2 33
Thickness of body at M_2 26
Symphysial height   35

195

(2) SH–7A–125

MANDIBLE

Howell (1969)

No mandibular dimensions were given in the original description.

TEETH

		Lower Teeth (Crown Dimensions)					
	I_1	I_2	C	PM_1	PM_2	M_1	M_2	M_3

Left	l	—	—	7·8	10·4	11·7*	c. 18·7*	c. 16·2*	c. 18·2*
Side	b	—	—	9·6	c. 17·5*	c. 18·9*	c. 16·8*	18·0*	14·8*
Right	l	—	?	?	11·2	?	c. 18·7*	c. 16·2*	c. 18·2*
Side	b	—	?	?	c. 17·5*	c. 18·9*	c. 16·8*	18·0*	14·8*

* Details of side are not given for these teeth thus the dimensions may refer to either or both sides.

SH–74A–21

MANDIBULAR FRAGMENT

No mandibular dimensions were given in the original description.

TEETH

		Lower Teeth (Crown Dimensions)						
	I_1	I_2	C	PM_1	PM_2	M_1	M_2	M_3

Right	l	—	—	8·8	c. 12·8	13·0	—	—	—
Side	b	—	—	9·7	—	13·75	—	—	—

SH–40–19

ULNA

Howell and Wood (1974)

Maximum length   315

(3) OMO 1

Day (1969)

SKULL

Maximum length   210

Maximum breadth   144*

Cranial index   68·5

The Omo Remains

TEETH
Right upper canine Length   8·9
                   Breadth   8·1
Left lower first molar Length   —
                       Breadth   11·5

POSTCRANIAL BONES
Provisionally the dimensions are within the range of modern man.

OMO II
SKULL
Maximum length   215   Cranial capacity 1,435±20
Maximum breadth   145
Cranial index   67·5

OMO III
No dimensions of value are available at present.

Originals (1) National Museum of Ethiopia, Addis Ababa, Ethiopia
(2) National Museum of Ethiopia, at present being studied at the Department of Anthropology, Berkeley Campus, University of California, U.S.A.
(3) National Museum of Ethiopia, Addis Ababa, Ethiopia.

Casts (1) Not available at present
(2) Not available at present
(3) Not available at present

References ARAMBOURG, C. and COPPENS, Y. 1967
Sur la découverte dans le Pléistocène Inferieur de la vallée de l'Omo (Éthiopie) d'une mandibule d'australopithécien. *C.R. Acad. Sci. Paris* 265, 589-590.
ARAMBOURG, C., CHAVAILLON, J. and COPPENS, Y., 1967
Expédition internationale de recherche paléontologiques dans la vallée de l'Omo (Éthiopie) en 1967. *Actes du 6è Congres Panafricain de préhistoire et d'Etudes du Quaternaire, Dakar*, 135-140.
ARAMBOURG, C. and COPPENS, Y. 1968
Découverte d'un australopithécien nouveau dans les gisements de l'Omo (Éthiopie). *S. Afr. J. Sci. 64*, 58-59.
DAY, M. H. 1969
Omo human skeletal remains. *Nature 222*, 1135-1138.

HOWELL, F. C. 1969
Remains of Hominidae from Pliocene/Pleistocene formations in the lower Omo basin, Ethiopia. *Nature 223*, 1234-1239.

HOWELL, F. C., FICHTER, L. S. and ECK, G. 1969
Vertebrate assemblages from the Usno Formation, White Sands and Brown Sands localities, lower Omo basin; Ethiopia. *Quaternaria 11*, 65-88.

BROWN, F. H. 1971
Radiometric dating of sedimentary formations in the lower Omo valley, southern Ethiopia. In *Calibration of hominoid evolution*, Eds. W. W. Bishop and J. A. Miller. Edinburgh: Scottish Academic Press.

BUTZER, K. W. 1971
Recent history of an Ethiopian delta. Chicago: Department of Geography; Chicago University.

COPPENS, Y., CHAVAILLON, J. and BEDEN, M. 1973
Résultats de la nouvelle mission de l'Omo (campagne 1972)— Découverte de restes d'Hominidés et d'une industrie sur éclats. *C.R. Acad. Sci. Paris 276*, 161-164.

HOWELL, F. C. and COPPENS, Y. 1974
Les faunes de mammifères fossiles de formations Plio/Plèistocènes de l'Omo en Éthiopie. *C.R. Acad. Sci. Paris 278*, 2275-2278.

MERRICK, H. V., HAESAERTS, P., DE HEINZELIN, J. and HOWELL, F. C. 1973
Artefactual occurrences associated with the Pliocene/Pleistocene Shungura Formation, southern Ethiopia. *Nature 242*, 572-575.

BUTZER, K. W. 1974
The Mursi, Nkalabong and Kibish formations, Lower Omo basin (Ethiopia). *Earliest man and environments in the Lake Rudolf basin: stratigraphy, paleoecology and evolution*. Chicago: Chicago University Press.

DE HEINZELIN, J. and HAESAERTS, P. 1974
Depositional history of the Shungura Formation. *Ibid*.

HOWELL, F. C. and WOOD, B. A. 1974
Early hominid ulna from the Omo basin, Ethiopia. *Nature 249*, 174-176.

BROWN, F. H. 1976
Isotopic ages and magnetostratigraphy of Omo Group Formations. In *Geological background to fossil man*. Ed. W. W. Bishop (in press).

The Hadar Remains

Synonyms and other names	1. aff. *Australopithecus robustus* 2. aff. *Australopithecus africanus* (*sensu stricto*) 3. aff. *Homo sp.* (Johanson and Taieb, 1976)
Site	The Hadar site is located in the Afar depression in the west central Afar sedimentary basin, near the Awash river about 300 km. north-east of Addis Ababa, Ethiopia.
Found by	The International Afar Research Expedition led by D. C. Johanson and M. Taieb, on 30th October, 1973. Further finds were made later that season and also in 1974 and 1975.
Geology	The Afar region is at the northern end of the East African rift valley and contains a Plio-Pleistocene sedimentary basin capped unconformably by an uppermost unit of Pleistocene gravels and sands. The Plio-Pleistocene strata have been termed the Central Afar Group and include a number of sequences, one of which has been referred to as the Hadar Formation. This Formation contains several members named, from below upwards, the Basal Member, the Sidi Hakoma Member, the Denen Dora Member and the Kada Hadar Member totalling some 140 m. of deposits. The sediments represent lacustrine, lake margin and associated fluvial deposits interspersed with volcanic tuffs and a single basalt flow. The hominid finds are in two principal groups, one about 40 m. above the basalt in the Sidi Hakoma Member and one extending to about 40 m. below the basalt (Taieb, Johanson, Coppens and Aronson, 1976).
Associated finds	Artefacts of many types have been recovered from the area, often in mixed assemblages although none from the hominid-bearing deposits. A few sites contained assemblages of an homogeneous character including one Acheulean site. The artefacts found in the area have been placed in five categories; (1) choppers, polyhedral and modified pebbles, crude flakes and protohandaxes; (2) bifaces and flakes of Middle Acheulean type; (3) flakes and cores of a Middle Stone Age Industry; (4) flakes, cores and waste of a Late Stone Age

The Hadar Remains

industry; (5) flakes, cores and waste from subrecent to recent industries (Corvinus, 1975).

Numerous fossil mammalian bones have been recovered from the site including proboscidians, hippopotamids, suids, giraffids, bovids, rodents, carnivores and primates, as well as numerous fossil reptilian and avian remains. Three faunal units have been suggested; a lower unit including Elephantidae associated with *Notochoerus euilus*, *Nyanzachoerus pattersoni*, *Tragelaphus cf. nakuae*, *Aepyceros sp.* and cercopithecines; a middle unit containing *N. euilus*, *T. cf. nakuae*, *Aepyceros sp.* and *Kobus*; and an upper unit containing *N. euilus*, *T. cf. nakuae*, *Kobus sp.*, Alcephalini and Elephantidae. These faunal assemblages may form the basis of palaeoecological and biostratigraphic interpretations in the future (Taieb, Johanson, Coppens and Aronson, 1976).

Dating The dating of the Hadar site is based upon stratigraphic, faunal, radiometric and palaeomagnetic data. The faunal correlation that fits most closely appears to be that with the Omo I Zone (Usno Formation and the Basal Member and Members A and B of the Shungura Formation). The time span given to this sequence in the Omo is c. 2·6–3·1 m.y. B.P. The potassium-argon measurements from the Hadar site have been performed upon the basalt and a tuff (SHT); of the five results published three from the tuff are inconsistent. Two of the results based on the basalt are in agreement at 2·9–3·0 m.y. B.P. with an experimental error of 200,000 years either way. The geomagnetic results show that the polarity of the basalt is reversed while the hominid fossil bearing sediments both above and below are normally polarized. It is suggested that the reversed sample represents the Kaena or the Mammoth event within the Gauss Normal Epoch (Taieb *et al.*, 1976).

Morphology (1) AL 211-1
The proximal end of a right femoral shaft lacking the head and part of the neck. The cross section of the neck is flattened and oval, the trochanteric fossa is well marked and the muscular markings are prominent.

200

AL 166–9

A temporal fragment that resembles material from both Swartkrans and East Rudolf.

(2) AL 288–1

An associated partial skeleton including a mandible, some cranial fragments, vertebrae, ribs, humeri, a right scapular fragment, ulnae, parts of radii, some hand bones, a sacrum, a left hip bone, a left femur, a right tibia and part of a fibula and a right talus. The pelvic remains suggest that the skeleton belonged to a female. The bones are remarkable for their small size, despite the fact that there is no evidence of immaturity; this indicates that the stature of this individual was diminutive.

Preliminary comparisons show similarities to material from the Sterkfontein site in the Transvaal.

AL 128–1, AL 129–1a–c

A group of material believed to be associated consisting of two proximal femoral fragments of opposite sides and a right distal femoral fragment as well as a right proximal tibial fragment. All of these remains resemble their counterparts in AL 288–1 (Johanson and Coppens, 1976).

(3) AL 199–1

A right maxillary fragment containing teeth that include the lateral incisor root, the canine, both premolars and the molars. This adult half-palate is shallow and has well developed alveolar prognathism. The maxillary sinus is large and the zygomatic root takes off above the distal portion of the first molar.

AL 200–1 a and b

A complete and undistorted palate with a full dentition. The tooth rows are sub-parallel with a broad anterior portion of the dental arcade to accommodate the spatulate central incisors. Marked diastemata are present between the lateral incisors and the canines. The palate is shallow with pronounced alveolar prognathism; the inferior nasal margin is guttered and the maxillary sinuses are large. The zygomatic roots are situated above the first molars. Marked similarities exist between the previous half-palate and this specimen, but 201 the former specimen is smaller in its general dimensions. This has led to the speculation that these two specimens represent male and female counterparts.

AL 266–1
A mandibular fragment consisting of the right side of the body, the symphysial region and part of the left side of the body containing the premolars and molars on the right but only the premolars and first two molars on the left. The dental arcade is rounded anteriorly and the tooth rows are straight diverging a little posteriorly. There is a moderately developed *planum alveolare* (Johanson and Taieb, 1976).

Dimensions (1) AL 166–9
Mandibular fossa width   25
Acoustic meatus length   30
                                        (Johanson and Coppens, 1976)

(2) AL 288–1
Mandible depth at M1   30
Mandible thickness at M1   19
Left Femur   Length 280★
Right Humerus   Length 235★
Humeral/Femoral Index   c. 83·9

(3) AL 199–1

| | Upper Teeth (Crown Dimensions) | | | | | | |
	I^1	I^2	C	PM^1	PM^2	M^1	M^2	M^3
Right l	—	—	8·7	7·3	7·1	10·1	11·7	11·3
side b	—	—	9·3	11·2	11·2	12·0	13·5	(12·7)

AL 200–1a

| | Upper Teeth (Crown Dimensions) | | | | | | |
	I^1	I^2	C	PM^1	PM^2	M^1	M^2	M^3
Right l	10·9	7·3	9·4	8·9	8·5	11·8	13·7	14·3
side b	8·5	7·0	11·0	12·2	12·1	13·2	15·0	15·0
Left l	10·8	7·4	9·4	9·0	8·5	11·8	13·8	14·2
side b	8·3	7·1	10·9	12·2	12·2	13·1	14·8	15·0

The Hadar Remains

AL 266-1

| | | Lower Teeth (Crown Dimensions) | | | | | |
		I_1	I_2	PM_1	PM_2	M_1	M_2	M_3
Right l	—	—	9·2	9·4	12·1	13·3	15·3
side b	—	—	10·1	10·8	12·0	14·0	13·7
Left l	—	—	9·1	8·9	12·1	—	—
side b	—	—	10·1	11·0	11·9	—	—

(Johanson and Taieb, 1976)

Affinities   At present the material from the Hadar site has only been announced (Taieb *et al.*, 1972, 1974 and 1975) and briefly described (Johanson and Taieb, 1976), thus the conclusions put forward are essentially preliminary and subject to revision following detailed study. With this in mind it has been suggested by the authors that there are the remains of three groups of hominids present at this site, one affined to the robust australopithecine material from Olduvai, Swartkrans and East Rudolf, one to the gracile australopithecine material from Sterkfontein and one to hominine material from Java and East Rudolf.

Originals   The National Museum of Ethiopia, Addis Ababa; at present being studied at the Cleveland Museum of Natural History, Cleveland, Ohio, U.S.A.

Casts   Not available at present.

References   TAIEB, M., COPPENS, Y., JOHANSON, D. C. and KALB, J. 1972
Dépôts sédimentaires et faunes du Plio-Pléistocène de la basse vallée de l'Awash (Afar central, Ethiopia). *C.R. Acad. Sci. Paris* *275*, 819–822.
TAIEB, M., JOHANSON, D. C., COPPENS, Y., BONNEFILLE, R. and KALB, J. 1974
Découverte d'Hominidés dans les séries Plio-Pléistocènes d'Hadar (Bassin de l'Awash; Afar, Ethiopia). *C.R. Acad. Sci. Paris* *279*, 735–738.
CORVINUS, G. 1975
Palaeolithic remains at the Hadar in the Afar region. *Nature* *256*, 468–471.

The Hadar Remains

TAIEB, M., JOHANSON, D. C. and COPPENS, Y. 1975
Expédition internationale de l'Afar, Éthiopie (3e campagne 1974);
découverte d'Hominidés Plio-Pléistocènes à Hadar. *C.R. Acad. Sci.
Paris 281,* 1297-1300.
TAIEB, M., JOHANSON, D. C., COPPENS, Y. and ARONSON, J. L. 1976
Geological and palaeontological background of Hadar hominid site,
Afar, Ethiopia. *Nature 260,* 289-293.
JOHANSON, D. C. and TAIEB, M. 1976
Plio-Pleistocene hominid discoveries in Hadar, Ethiopia. *Nature 260,*
293-297.
JOHANSON, D. C. and COPPENS, Y. 1976
A preliminary anatomical diagnosis of the first Plio-Pleistocene
hominid discovered in the Central Afar, Ethiopia. *Am. J. Phys.
Anthrop. 45,* 217-233.

Southern Africa

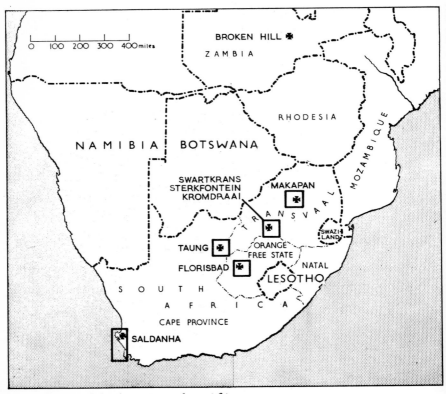

Fig. 67 Hominid fossil sites in southern Africa

The Rhodesian Remains

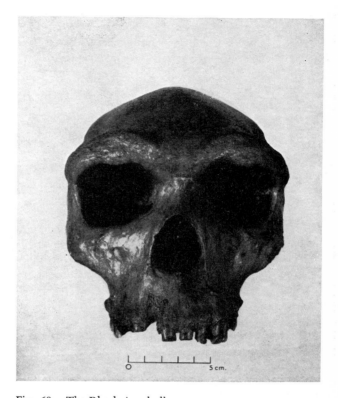

Fig. 68   The Rhodesian skull
Frontal view
Photographed by courtesy of the Trustees of the British Museum (Nat. Hist.)

Synonyms and other names   *Homo rhodesiensis* (Woodward, 1921); *Cyphanthropus rhodesiensis* (Pycraft, 1928); *Homo sapiens rhodesiensis* (Campbell, 1964)
Rhodesian man; Broken Hill man; Kabwe man

Site   The Broken Hill Mine, Broken Hill, Zambia.

Found by   T. Zwigelaar, 17th June, 1921, cranium; other remains found by A. S. Armstrong, 1921, A. W. Whittington, 1921 and H. Hrdlička, 1925.

207

The Rhodesian Remains

Geology The mine included two kopjes or small hills of dolomitic limestone which contained lead and zinc ore. One of the hills was tunnelled at its base by a cave filled with fossilized and mineralized bones. During the clearance of this cavern the skull was found at its farthest and deepest point about 60 feet below ground level. Subsequent excavations produced the rest of the remains, but continued mining has destroyed the original cave.

Associated finds With the hominid bones, although not associated on a living floor, were some quartz and chert stone tools. These implements belong to African flake cultures known as the Stillbay and Proto-Stillbay of the Middle Stone Age. In addition there were several bolas stones and a few bone tools.
The associated fauna included fossil birds, reptiles and mammals, many of which belong to living species. A recent mammalian faunal list for the Broken Hill site has been given by Cooke (1964); forms identified include a large primate (?*Simopithecus sp.*), mongoose (*Herpestes ichneumon*), large carnivores (*Panthera leo*, *Panthera pardus*), an extinct carnivore (*Leptailurus hintoni*), African elephant (*Loxodonta africana*), zebra (*Equus burchelli*), black rhinoceros (*Diceros bicornis*) and several artiodactyls including an extinct buffalo (*Homoioceras bainii*).

Dating Initially it was believed that the skull and the other bones may be of different ages because of variations in the mineral content of the specimens. Oakley (1947) suggested that these differences may be due to local variations in the mineral constituents of the soil at the site of burial. Further chemical and radiometric investigations indicate that the skull and the other bones are ancient and of the same age (Oakley, 1957, 1958). A new excavation at Broken Hill has produced more stone implements of a similar type to those found with the skeletal remains. The new finds have been attributed to the early Middle Stone Age (Upper Pleistocene) after comparison with the tools from another Rhodesian site which contains an established sequence (Clark, 1947, 1959). Thus on faunal, archaeological and chemical grounds it seems that Rhodesian man lived during the Upper Pleistocene, possibly about 40,000 years B.P.

The Rhodesian Remains

Recently it has been suggested that the dating of the Rhodesian site should be revised on the grounds of a revaluation of the artefacts and the fauna. The new date has been given a minimum age of 125,000 years B.P. (Klein, 1973).

Morphology The remains belong to at least three and possibly four individuals and consist of a skull, a parietal, a maxilla, a humerus, a sacrum, two ilia, several femoral fragments and two tibiae.

THE SKULL
The cranium is heavily built with massive brow ridges, a retreating forehead and a flattened vault; the occipital region is rounded above the occipital torus but flattened beneath. The foramen magnum is placed well forward and faces downward, indicating an erect head carriage. The mastoid process is of moderate size but the supra-mastoid crest is prominent. The greatest diameter of the cranium is situated very low.

Fig. 69   The Rhodesian skull
Left lateral view
Photographed by courtesy of the Trustees of the British Museum (Nat. Hist.)

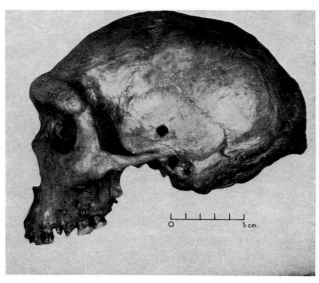

209

The Rhodesian Remains

The face is very long with inflated, but flat, maxillae having no canine fossae; the lateral walls of the nose pass smoothly on to the face but there is a nasal spine at the apex of two ridges which lead back to join the lateral wall half-way up the nasal opening. The alveolar processes of the maxillae are very deep and the bony palate is extremely large both in width and length.

The separate maxilla was re-examined by Wells (1947) who found that it differs in several respects from the corresponding bone in the skull. It is a smaller bone with a transversely arranged zygomatic process and a canine fossa 'modelling essentially as in modern human skulls'. However, he concluded that the maxilla and the skull belonged to a single type, perhaps not even specifically distinct from *Homo sapiens*.

Fig. 70   The Rhodesian skull
Basal view
Photographed by courtesy of the Trustees of the British Museum (Nat. Hist.)

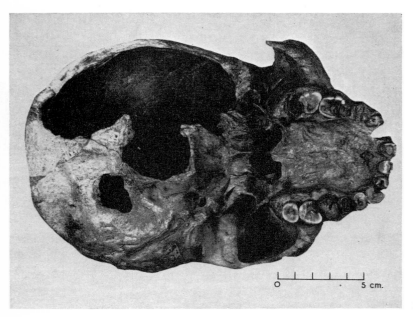

0   5 cm.

The Rhodesian Remains

THE TEETH

The teeth are large and set in a horseshoe-shaped arcade. All the teeth are considerably worn and most are affected by caries. The crowns are generally of modern form and the third molar is reduced in size.

THE POST-CRANIAL BONES

These bones have few, if any, features which lie outside the range of normal variation of modern man. The limb bones are stout and long, indicating tall stature, whilst the lower limb bones have no features which are incompatible with an upright stance and a bipedal striding gait.

A new femoral shaft fragment has recently come to light after being in private possession for many years (Clark *et al.*, 1968). The new fragment does not fit with any of the previously known femoral fragments, thus increasing the possible number of individuals found at this site.

Dimensions SKULL

Morant (1928)

Max. Length 208·5   Max. Breadth 144·5

Cranial Index 69·3 (Dolichocephalic)

Cranial Capacity 1,280 cc

POST-CRANIAL BONES

Pycraft (1928)

Sacrum Length 105   Sacrum Breadth 110

		Right	*Left*
Femora:	Max. Diameter of Heads	52·0	50·0
	Ant./Post. Diameter of Shaft	30·0	27·5
Humerus:	Ant./Post. Diameter of Shaft	20·0	—

TEETH

Pycraft (1928)

		Upper Teeth (Crown Dimensions)						
		I1	*I2*	*C*	*PM1*	*PM2*	*M1*	*M2*	*M3*
Left	l	8·0	7·0	10·0	7·5	—	—	13·0	9·0
Side	b	8·5	8·0	11·0	11·0	—	—	14·0	12·0
Right	l	8·0	—	—	—	—	(14·0)	12·5	—
Side	b	8·5	—	—	—	—	13·5	13·5	—

( ) Estimated

The Rhodesian Remains

Affinities The Rhodesian cranium has several points of similarity with those of European Neanderthal man whilst remaining distinct in the structure of the post-cranial bones, a situation which raised doubts about the correctness of associating the type skull with the remainder of the skeleton. These doubts were reinforced by the rather unsatisfactory circumstances of the find. However, the chemical and radiometric evidence seems to have established the contemporaneity of the remains.

The resemblances of the Rhodesian skeletons to that of Neanderthal man did not escape Woodward (1921) but the position of the foramen magnum being much more modern in the Rhodesian cranium than in the Neanderthalers available to him, he felt obliged to create a new species of man, *Homo rhodesiensis*. Pycraft (1928) was convinced of the peculiarity of the stance and gait of this form and created a new genus, *Cyphanthropus rhodesiensis* or 'Stooping man', a concept which gained little support since Le Gros Clark (1928) explained the error of interpretation which led to this viewpoint.

Morant (1928) has shown that the skull can be distinguished from those of Neanderthal man by a number of metrical characteristics and that it tends to resemble *Homo sapiens*; however, he concluded that Rhodesian man and Neanderthal man seem more closely related to each other than either is to *Homo sapiens* and also that they are equally related to all races of *Homo sapiens*.

Broken Hill man was placed near the point of divergence of Neanderthal and modern man by von Bonin (1928–1930) on the basis of further comparative skull measurements.

The discovery of the Saldanha calvarium (*q.v.*) is particularly significant at this point. In its form and dimensions it closely resembles the Rhodesian skull, confirming that this is not an isolated or aberrant specimen. Singer (1954) regarded the Rhodesian and Saldanha people as African Neanderthalians, unlike the European but similar to the Asiatic representatives of this group (Solo man).

Howell (1957) has denied Neanderthal penetration 'south of the Sahara' and regards Rhodesian man, as well as other related southern African forms, as racially distinct. Recently, in a review of the position of Rhodesian man, Wells (1957) suggests that the primitive pithecanthropine stock gave rise to 212

The Rhodesian Remains

a basic type of *Homo sapiens* which became widely dispersed and underwent regional differentiation into a number of offshoots represented by Broken Hill, Neanderthal and Solo man—these three lines becoming extremely specialized and then extinct.

Coon (1963) seeks to show that Rhodesian and Saldanha man are both forms of *Homo erectus* leading towards a possible negro evolutionary line, as a part of his general polyphyletic theory of racial origin. This view has received little support and much criticism. An anatomical comparison of the Rhodesian and Petralona skulls shows that they share some similarities (Murrill, 1975).

It is clear that the principal resemblances of the Rhodesian skull are to the Saldanha specimen, but the classification of these remains is still controversial. It would be widely agreed that Rhodesian man is a member of the genus *Homo* but the species to which he belongs remains in dispute. However, opinion is growing in favour of classifying Neanderthal and Neanderthaloid forms as sub-species of *Homo sapiens* (Campbell, 1964).

Originals   British Museum (Natural History), Cromwell Road, South Kensington, London, S.W.7.

Casts   The University Museum, University of Pennsylvania, Philadelphia 4, Pennsylvania, U.S.A.

References   WOODWARD, A. S. 1921
A new cave man from Rhodesia, South Africa. *Nature 108*, 371-372.
PYCRAFT, W. P. *et al.* 1928
Rhodesia man and associated remains. Ed. F. A. Bather, London: British Museum (Natural History).
CLARK, W. E. LE GROS 1928
Rhodesian man. *Man 28*, 206-207.
MORANT, G. M. 1928
Studies of Paleolithic Man III. The Rhodesian skull and its relations to Neanderthaloid and modern types. *Ann. Eugen. 3*, 337-360.
BONIN, G. VON 1928-1930
Studien zum *Homo rhodesiensis*. *Z. Morph. Anthr. 2*, 347-381.
CLARK, J. D. *et al.* 1947
New studies on Rhodesian man. *J. R. Anthrop. Inst. 77*, 7-32.

The Rhodesian Remains

WELLS, L. H. 1947
In *New studies on Rhodesian man*. J. D. Clark *et al*. II. A note on the broken maxillary fragment from the Broken Hill cave. *J. R. Anthrop. Inst. 77*, 11-12.

SINGER, R. 1954
The Saldanha skull from Hopefield, South Africa. *Am. J. Phys. Anthrop. 12*, 345-362.

OAKLEY, K. P. 1957
The dating of the Broken Hill, Florisbad and Saldanha skulls. *Proc. Third Pan-African Cong. Prehist., Livingstone, 1955*, 76-79. Ed. J. D. Clark. London: Chatto and Windus.

WELLS, D. H. 1957
The place of the Broken Hill skull among human types. *Ibid.*, 172-174.

OAKLEY, K. P. 1958
The dating of Broken Hill (Rhodesian man). In *Hundert Jahre Neanderthaler*, 265-266. Ed. G. H. R. von Koenigswald, Utrecht: Kemink en Zoon.

CLARK, J. D. 1959
Further excavations at Broken Hill, Northern Rhodesia. *J. R. Anthrop. Inst. 89*, 201-231.

LEAKEY, L. S. B. 1959
A preliminary re-assessment of the fossil fauna from Broken Hill, N. Rhodesia. *J. R. Anthrop. Inst. 89*, 225-231.

COON, C. S. 1963
The origin of races, 621-627. London: Jonathan Cape.

CAMPBELL, B. 1964
Quantitative taxonomy and human evolution. In *Classification and human evolution*, 50-74. Ed. S. L. Washburn. London: Methuen and Co. Ltd.

COOKE, H. B. S. 1964
Pleistocene mammal faunas of Africa, with particular reference to Southern Africa. In *African ecology and human evolution* 65-116. Eds. F. C. Howell and F. Bourlière. London: Methuen and Co. Ltd.

CLARK, J. D., BROTHWELL, D. R., POWERS, R. and OAKLEY, K. P. 1968
Rhodesian man: notes on a new femur fragment. *Man 3*, 105-111.

KLEIN, R. G. 1973
Geological antiquity of Rhodesian man. *Nature 244*, 311-312.

MURRILL, R. I. 1975
Z. Morph. Anthr. 66 (2), 176-187.

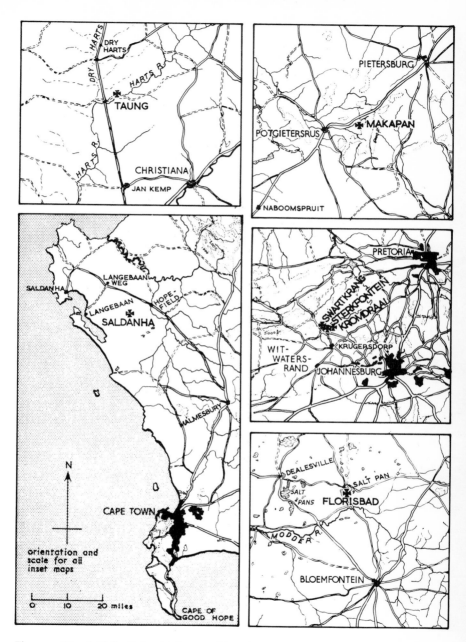

Fig. 71 Hominid fossil sites in the Republic of South Africa.

The Taung Skull

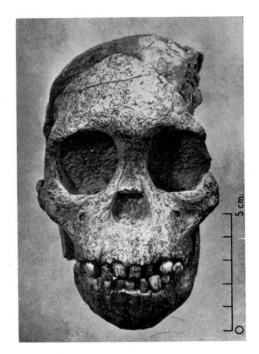

Fig. 72   The Taung skull
Frontal view
*Courtesy of Professor P. V. Tobias,
photographed by A. R. Hughes*

Synonyms *Australopithecus africanus* (Dart, 1925)
and other names *Homo transvaalensis* (Mayr, 1950); *Australopithecus africanus
africanus* (Robinson, 1954)
'Ape-man'; 'Man-ape'; 'Near-man'

Site A limestone quarry six miles south-west of Taung railway
station, 80 miles north of Kimberley, Republic of South
Africa.

Found by A quarryman, M. de Bruyn; recognized by R. A. Dart,
November, 1924.

Geology The skull was found in sandy breccia which formed the filling   216

of a cave cut into the face of a dolomitic limestone escarpment. Mining of the tuffaceous limestone, for the manufac-
ture of cement, led to the discovery after a blasting operation.
Formerly the Taung cave deposit was believed to be part of
the Thabaseek Travertine (Peabody, 1954). Recent investigations have suggested that the skull was associated with the
younger Norlim Tufa, in a cave filling intrusive in the
Thabaseek formation. Electron microscopic examination of
the matrix adhering to the skull suggests its association with a
'wet' depositional phase in contrast to most of the Taung
fauna which comes from 'dry-phase' deposits (Butzer, 1974a).

Associated finds No artefacts have been recovered from the cave deposit. The
fossil mammalian fauna found with the skull includes three
insectivores (*Elephantulus cf. brachyrhynchus*, *Mylomygale
spiersi*, *Crocidura taungensis*), a bat (*Rhinolophus cf. capensis*),
several baboons (*Parapapio antiquus*, *P. jonesi*, *P. whitei*,
P. izodi, *P. wellsi*), numerous rodents (*Thallomys debruyni*,
Gypsorhynchus darti, *G. minor*, *Cryptomys robertsi*, *Pedetes
gracile*, *Dendromus antiquus*, *Protomys campbelli*, *Petromus minor*),
two hyraces (*Procavia capensis*, *P. transvaalensis*) and two artiodactyls (*Cephalophus parvus*, *Oreotragus longiceps*).
The bones of neither carnivora nor proboscidea have been
positively identified in the deposit (Cooke, 1963).

Dating Accurate dating of the remains from Taung has proved
difficult since material from this type of deposit is not amen-
able to the chemical and radiometric techniques that are
available at present. Kurtén (1962) attributed the Taung skull
to the First Interglacial (Günz-Mindel or Antepenultimate
Interglacial) on the grounds of faunal correlation. The balance
of evidence from geological and faunal studies suggested that
the Taung deposit was laid down early in the sequence of
australopithecine sites, during the Upper Villa-franchian part
of the Lower Pleistocene (Ewer 1957; Oakley, 1954, 1957,
1964).
    Cooke (1970) attempted to correlate some of the South
African hominid sites with those that are better dated
from East Africa. He concluded that the Sterkfontein and
Makapansgat (*q.v.*) sites are 2·5–3·0 million years old. Wells 217

(1967) has expressed doubts, on faunal grounds, that Taung is the oldest of these sites and has been supported by Freedman's (1970) findings in relation to the cercopithecoids.

Recently it has been suggested, on the basis of a geomorphological dating method, that the Taung site might be the youngest of all the Transvaal cave sites, the date of opening of the Taung caves being given as 0·87 million years B.P. (Partridge, 1973). While accepting that the Taung hominid 'clearly postdates the gracile australopithecines from Sterkfontein and Makapansgat', on the basis of his own studies, Butzer (1974b) does not accept Partridge's attempt at dating the site as valid.

Morphology THE SKULL

The specimen consists of the greater part of a juvenile skull which contains a remarkable endocast of the brain. The facial skeleton is intact and the dentition complete. Most of the base of the skull is preserved but much of the vault is missing. The lower parts of the body and the angles of the mandible are broken.

The cranium appears to be globular with neither frontal flattening nor supra-orbital torus formation, but the glabella is prominent; the sphenoid and parietal bones appear to have been in contact in the temporal fossa. The foramen magnum is set well forward beneath the skull. The face is undistorted, having large rounded orbits, flattened nasal bones and a square nasal opening which runs without interruption on to the maxillae; there is no nasal spine. The nasal flattening gives the face a 'dished' appearance which is accentuated by the degree of sub-nasal prognathism of the maxillae.

The dental arcade is regular and parabolic enclosing a shallow palatal vault marked by an incisive foramen; the palatine foramina are set behind a line drawn posterior to the first permanent molars.

MANDIBLE

This bone is represented by the body and the alveolar processes of bone with the teeth in place. The angles of the mandible and the rami are largely absent. The symphysial region slopes backwards, there being neither chin nor simian shelf   218

The Taung Skull

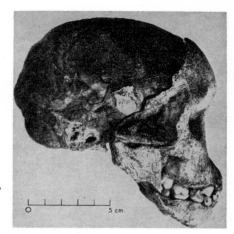

Fig. 73   The Taung skull
Right lateral view
Courtesy of Professor P. V. Tobias,
photographed by A. R. Hughes

present. The body of the mandible is thickened particularly in the region of the erupting first molar.

TEETH

The milk dentition is complete in both upper and lower jaws, and the upper and lower first permanent molars are in process of eruption.

The Upper Deciduous Incisors are heavily worn and damaged. The teeth are well separated one from another, a feature characteristic of other hominids at this stage of dental development.

The Upper Deciduous Canines are also worn and damaged but appear to be small and spatulate.

The Upper First Deciduous Molars are considerably worn particularly on the lingual half of the crowns, but the mesio-buccal angles of the teeth seem exaggerated in their development.

The Upper Second Deciduous Molars are also worn but enough of the crown morphology is discernible to show that they resemble the first milk molars.

The Upper First Permanent Molars are both perfect. The crowns bear four main cusps and show anterior foveae as well as a Carabelli complex of grooves. These features are characteristic of teeth from Sterkfontein but not of those from Swartkrans.

The Lower Deciduous Incisors are heavily worn and damaged.   219

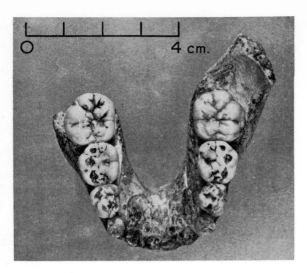

Fig. 74   The Taung mandible
          Occlusal view
          Courtesy of Professor P. V. Tobias, photographed by
          A. R. Hughes

The left Lower Deciduous Canine is well preserved and has a distal cusplet. The tooth does not appear to project appreciably above the occlusal line.

The Lower First Deciduous Molars are little damaged and resemble the equivalent teeth from Sterkfontein in some detail: five main cusps are present and the Taung specimens have small sixth cusps.

The Lower Second Deciduous Molars are appreciably worn but the typical dryopithecine cusp pattern can still be defined. Well developed sixth cusps have obliterated the posterior foveae but the anterior foveae are prominent.

The Lower First Permanent Molars are unworn and display five principal cusps, sixth cusps have obliterated the posterior foveae. Deeply incised buccal grooves are present, the anterior ones ending in a distinct pit.

Dimensions   SKULL
              Max. Length (Glabella/Inion) 127·0   Cranial Index 62·4
              Cranial Capacity 405 cc (440 cc adult
              value including 8% for growth)
                          Holloway (1970)   220

The Taung Skull

TEETH

		Upper Dentition	(Crown Dimensions)			
		DI1	DI2	DC	DM1	DM2	M1 (Perm.)

		DI1	DI2	DC	DM1	DM2	M1 (Perm.)
Left	l	—★	—★	—★	8·8	—★	12·75
side	b	—★	—★	6·0	10·0	—★	14·0
Right	l	—★	—★	6·8	8·8	10·1	12·75
side	b	—★	—★	5·8	10·0	11·0	14·0

		Lower Dentition (Crown Dimensions)					
		DI1	DI2	DC	DM1	DM2	M1 (Perm.)
Left	l	—★	—★	6·5	—★	11·5	14·0
side	b	—★	—★	5·3	—★	10·6	13·5
Right	l	—★	—★	—★	8·7	11·5	14·0
side	b	—★	—★	—★	8·0	10·7	13·5

★ Damaged

Affinities The affinities of the Taung skull will be discussed with the rest of the australopithecine remains.

Original Department of Anatomy, Medical School, University of the Witwatersrand, Johannesburg, Republic of South Africa.

Casts The University Museum, University of Pennsylvania, Philadelphia 4, Pennsylvania, U.S.A.

References DART, R. A. 1925
Australopithecus africanus: the man-ape of South Africa. *Nature 115,* 195–199.
DART, R. A. 1926
Taungs and its significance. *Nat. Hist. 3,* 315–327.
DART, R. A. 1934
The dentition of *Australopithecus africanus. Folio anat. jap. 12,* 207–221.
BROOM, R., and SCHEPERS, G. W. H. 1946
The South African fossil ape-men, the *Australopithecinae. Transv. Mus. Mem. 2,* 1–272.
MAYR, E. 1950
Taxonomic categories in fossil hominids. *Cold Spring Harbour Symposia on Quantitative Biology 15, 109–118.*
OAKLEY, K. P. 1954
The dating of the *Australopithecinae* of Africa. *Am. J. Phys. Anthrop. 12,* 9–28.

PEABODY, F. E. 1954
Travertines and cave deposits of the Kaap escarpment of South Africa and the type locality of *Australopithecus africanus* Dart. *Bull. geol. Soc. Am.* *65*, 671-705.

ROBINSON, J. T. 1954
The genera and species of the *Australopithecinae. Am. J. Phys. Anthrop.* *12*, 181-200.

ROBINSON, J. T. 1956
The detention of the *Australopithecinae. Transv. Mus. Mem.* *9*, 1-179.

EWER, R. F. 1957
Faunal evidence on the dating of the *Australopithecinae. Proc. Third Pan-African Cong. Prehist., Livingstone, 1955*, 135-142. Ed. J. D. Clark. London: Chatto and Windus.

OAKLEY, K. P. 1957
Dating the australopithecines. *Ibid.*, pp. 155-157.

KURTÉN, B. 1962
The relative ages of the australopithecines of Transvaal and the pithecanthropines of Java. In *Evolution und Hominisation*, 74-80. Ed. G. Kurth. Stuttgart: Gustav Fischer Verlag.

COOKE, H. B. S. 1964
Pleistocene mammal faunas of Africa, with particular reference to South Africa. In *African ecology and human evolution*, 65-116. Eds. F. C. Howell and F. Bourlière. London: Methuen and Co. Ltd.

OAKLEY, K. P. 1964
Frameworks for dating fossil man. London: Weidenfeld and Nicolson.

WELLS, L. H. 1967
In *Background to evolution in Africa*, 105-106. Eds. W. W. Bishop, and J. D. Clark. Chicago: Chicago University Press.

COOKE, H. B. S. 1970
Notes from members: Dalhousie University, Halifax, Canada. *News Bull. Soc. Vertebra. Palaeont.* *90*, 2.

HOLLOWAY, R. L. 1970
New endocranial values for the australopithecines. *Nature*, *227*, 199-200.

FREEDMAN, L. 1970
A new checklist of fossil Cercopithecoidea of South Africa. *Palaeont. Afr.* *13*, 109-110.

PARTRIDGE, T. C. 1973
Geomorphological dating of cave openings at Makapansgat, Sterkfontein, Swartkrans and Taung. *Nature 246*, 75-79.

BUTZER, K. W. 1974a
Palaeoecology of South African australopithecines: Taung revisited. *Curr. Anthrop. 15*, 367-382.

BUTZER, K. W. 1974b
Comment—reply. *Curr. Anthrop. 15*, 413-416.

The Sterkfontein Remains

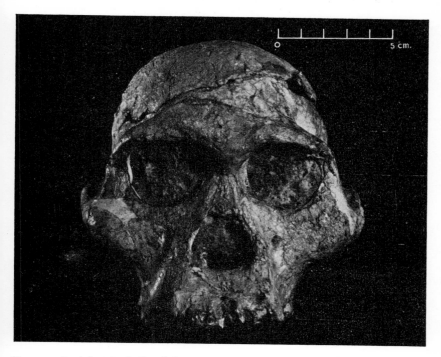

Fig. 75   Sterkfontein skull 5 ('Plesianthropus 5')
Frontal view
Courtesy of Professor J. T. Robinson

Synonyms
and other names
Australopithecus transvaalensis (Broom, 1936); *Plesianthropus transvaalensis* (Broom, 1937); *Homo transvaalensis* (Mayr, 1954); *Australopithecus africanus transvaalensis* (Robinson, 1954); *Homo africanus* (Robinson, 1972); 'Ape-man'; 'Man-ape'; 'Near-man'.

Site   Sterkfontein Type Site and Sterkfontein Extension Site, seven miles north-west of Krugersdorp, near Johannesburg, Transvaal, Republic of South Africa.

Found by   1. R. Broom, August, 1936 (Cranial and post-cranial bones, teeth).

2. R. Broom and J. T. Robinson, 1947–1948 (Cranial and post-cranial bones, teeth).

223

The Sterkfontein Remains

The majority of the hominid material is derived from the Type Site.

Geology At Sterkfontein a number of caves honeycomb a Pre-Cambrian formation of impure dolomitic limestone. The cave fillings consist of calcareous bone breccia that has been mined for many years and burned in kilns for lime. The principal site is the remains of a large cavern which communicates with the surface; gradually it had become filled with debris and bones until the cavern floor collapsed into an underlying cavity in the rock. The process then recurred, the new cavern again filling from above. Because of this mode of formation, the breccia is not of uniform character, neither is it regularly stratified. The geology of Sterkfontein has been investigated in detail (Brain, 1957, 1958; Robinson and Mason, 1962) and further work has been reported more recently (Tobias and Hughes, 1969; Tobias, 1973). This has shown that there is an earlier 'bone-bearing breccia' under the travertine that was formerly believed to be the floor of the main deposit.

Associated finds A number of stone tools have been recovered from the Extension Site (Robinson, 1957; Brain, 1958; Robinson and Mason, 1962). They comprise two hand-axes, choppers, flakes, irregular artefacts and a spheroid. The tools are made of diabase, quartzite or chert. The culture has been described as Late Oldowan or Early Chelles-Acheul in character, and has been attributed to later pithecanthropines who may have occupied the site (Robinson and Mason, 1962).
The fossil mammalian fauna recovered from Sterkfontein Type Site include insectivores (*Chlorotalpa spelaea, Elephantulus langi, Crocidura cf. bicolor, Suncus etruscus, Myosorex robinsoni*), primates (*Parapapio jonesi, P. broomi, P. whitei, Cercopithecoides williamsi*), numerous rodents, some carnivores (*Canis mesomelas pappos, Canis brevirostris, Lycyaena silbergi, Therailurus barlowi, Megantereon gracile*), hyraces (*Procavia antiqua, P. transvaalensis*) and some artiodactyls ('*Tapinochoerus*' meadowsi, *Hippotragus broomi, Gazella wellsi*) (Cooke, 1964). The presence of an equid (*Equus*) at the Extension Site has been reported (Robinson, 1958).
Later Cooke (1970) has attempted to correlate, on faunal grounds, the Sterkfontein site with those that are better dated 224

in East Africa. He concluded that the Sterkfontein site should be dated at 2·5–3·0 million years B.P., about twice as old as had been formerly believed. Vrba (1975) has studied bovid material from Sterkfontein, Swartkrans and Makapansgat and concludes that the Type Site (Sts) should be dated at 1·75–2·5 million years B.P. with heavier bush cover than the Extension Site (SE) dated at about 0·6 million years B.P. with less bush cover.

Dating The same difficulties arise in dating the Sterkfontein sites as were encountered at the other australopithecine sites; the problem has been discussed by a number of authors (Howell, 1955; Ewer, 1957; Oakley, 1954, 1957; Brain, 1958; Robinson and Mason, 1962). Kurtén (1962) attributed the Sterkfontein remains to the First Interglacial (Günz–Mindel or Antepenultimate Interglacial) on the grounds of faunal correlation. The somewhat conflicting faunal evidence for the dating of the site (see above) has been added to by a recent attempt to date the opening of the Sterkfontein caves by a geomorphological method (Partridge, 1973). The conclusion of this study is that the Sterkfontein cave opened as early as 3·26 million years B.P. and began to collect the hominid-bearing deposit. The validity of this method has been seriously questioned (Butzer, 1974).

No radiometric methods, palaeomagnetic determinations or biochemical dating techniques have proved possible at Sterkfontein.

Morphology The earlier excavation produced a broken cranium, maxillary, zygomatic and nasal bones, some mandibular fragments, socketed and isolated teeth, the lower end of a left femur and a capitate. The later excavation was rewarded by an almost complete cranium, several other damaged and incomplete crania, a nearly complete mandible, parts of other mandibles, several maxillae, numerous socketed and isolated teeth, part of a scapula, the upper end of a humerus, vertebrae, ribs, right and left innominate bones, the proximal end of a femur and the distal end of a femur.

Subsequently further material has come to light including a complete dentition, a sacrum and a second femoral fragment, almost identical to the first but of the opposite side. Recent

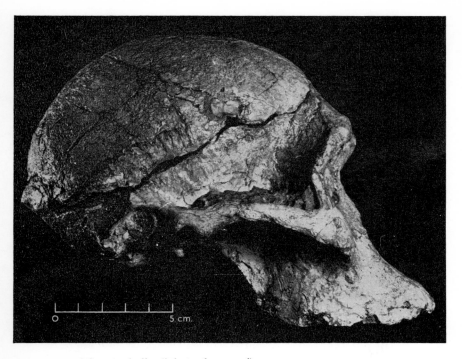

Fig. 76   Sterkfontein skull 5 ('Plesianthropus 5')
Right lateral view
Courtesy of Professor J. T. Robinson

excavations (1968–1974) have produced 21 isolated teeth,
four adult articulated lumbar vertebrae (StW/8), an adult
cranium (StW/12, 13 and 17), a young adult mandible with
eight teeth (StW/14) and an adolescent right maxilla with five
teeth (StW/18a, b and c) (Tobias, 1973).

SKULL

The best-preserved cranium, Skull 5, is that of a mature
female widely known colloquially as 'Mrs Ples'; it is virtually
complete, lacking only the upper teeth. The vault is rounded
and marked anteriorly by a modest supra-orbital ridge and
some prominence of the glabella. The occipital crest is weak
and the nuchal plane low. The foramen magnum is set well
beneath the cranium and the mastoid processes are small. The
greatest breadth of the cranium is bitemporal.

226

Even more recently a new fossil skull (StW/53) has been recovered from the West Pit at Sterkfontein (Hughes and Tobias, 1977). It consists of a cranio-facial fragment, including some teeth, and has been affined to *Homo habilis* known from Olduvai Gorge (*q.v.*).

MANDIBLE

The more complete mandible (Sts 52b) is rather crushed, particularly on the right side; most of the teeth are preserved but they are heavily worn. The jaw is large and robust, the symphysial region is well preserved and there is neither chin nor simian shelf. The body of the mandible is stout and the rami are tall.

PERMANENT TEETH

About 150 permanent teeth are known from Sterkfontein.

The Upper Incisors are small and moderately shovelled, having marginal ridges on their lingual faces.

The Upper Canines are symmetrical and pointed, projecting a little beyond the adjacent teeth; the lingual face of these teeth have parallel grooves whilst two specimens have small lingual tubercles.

The Upper First Premolars are bicuspid and have well defined buccal grooves. The roots are poorly developed.

The Upper Second Premolars are all worn but resemble the first premolars in having the same occlusal features and buccal grooves. The roots tend to be narrow.

The Upper First Molars are rhomboid in shape, having a simple quadrituberculate cusp pattern. A Carabelli complex seems to be a constant feature of these teeth.

The Upper Second Molars are similar to the first molars but slightly larger. Characteristically a well developed cingulum is present running from the lingual groove to the lingual end of the mesial face. An extra cusp is often found distally.

The Upper Third Molars are essentially the same size and shape as the second molars, having a lingual groove and part of a Carabelli complex. The fissure pattern of the occlusal surface is complicated.

⋆   ⋆   ⋆

The Lower Incisors tend to be shovelled and have horizontal incisal margins. Both central and lateral specimens have five well developed mamelons.

The Lower Canines differ considerably from the upper in that the crown is always asymmetrical; the apex of the tooth is distal to its midline and the cingulum reaches higher up the mesial face than the distal face. The cingulum on the distal face forms a distinct cusplet.

The Lower First and Second Premolar crowns are bicuspid and asymmetrical. The lingual cusp is smaller than the buccal cusp, and the anterior and posterior foveae are well defined.

The Lower First Molars are rectangular having five cusps and a larger anterior fovea.

The Lower Second Molars have six cusps and a moderate anterior fovea.

The Lower Third Molars tend to be triangular; they usually have six cusps and a distinct anterior fovea.

DECIDUOUS TEETH

The Upper First Molar is asymmetrical due to the large mesio-buccal angle of its crown; a fifth cusp is present.

The Upper Second Molar is similar in form to the first permanent molar and has a distinct Carabelli cusp.

★   ★   ★

The Lower Incisors are either damaged or worn.

The Lower Canine has a moderately high crown with distal and mesial cusplets.

The Lower First Molar has five cusps and a large anterior fovea.

The Lower Second Molar has a similar cusp arrangement to the first permanent molar.

THE POST-CRANIAL BONES

The Scapula (Sts 7) is a little crushed and has lost its lower and inner half. The neck of the bone is not clearly defined and there is no scapular notch.

The Humerus (Sts 7) belongs to the same individual as the scapula and has the head and upper end of the shaft in good condition. The remainder of the shaft is badly crushed, and the lower end is missing. The head of the bone is very like that of modern man, having greater and lesser tuberosities separated by an intertubercular groove and a well rounded articular surface.

Of the two *Hip Bones* (Sts 14) the right one is almost complete, lacking only its anterior superior iliac spine. The ilium is broad and similar to that of man but its ischium and pubis are

The Sterkfontein Remains

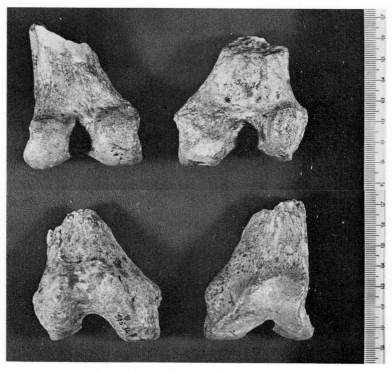

Fig. 77   Femoral fragments from Sterkfontein
Above: TM 1513 and Sts 34 Posterior views
Below: Sts 34 and Tm 1513 Anterior views
Photographed by courtesy of the Director of the Transvaal Museum, Pretoria

badly crushed. The iliac pillar, which in man extends from the acetabulum to the tubercle of the crest, is feeble and runs forward to the anterior superior iliac spine. The sacral articulation is small, as is the roughened region above for the sacro-iliac ligaments. The ischial tuberosity, which is irregular and flattened, is set well away from the edge of the acetabulum to produce a strikingly long pelvic ischial segment. The orientation of the pelvis is distinctive; in man the iliac crests are mainly directed forwards, but in this form they were mainly directed

229

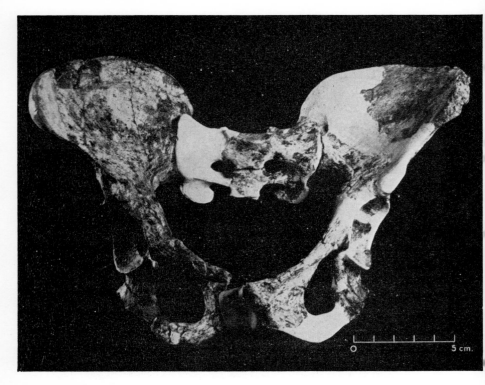

Fig. 78   The Sterkfontein pelvis (restored)
            Courtesy of Professor J. T. Robinson

laterally. The left hip bone is badly damaged and has been
much repaired. Some doubts have been cast on the correctness
of its reconstruction (Day, 1974).
Closely associated with the pelvis was part of a left *Femur*. The
specimen is almost valueless since it lacks a head and the shaft
consists of irregularly glued fragments that are barely
diagnostic. Two other fragments, the lower ends of right and
left femora, are very similar in size and shape. They closely
resemble the femur of man but differ in the relative depth and
forward extension of their intercondylar grooves.
The Capitate Bone (Tm 1526) and a *Phalangeal Fragment* were
recovered during the early excavation. The capitate is small
and similar to that of man in its essential features. The post-   230

cranial material from this site has been analysed recently and conclusions reached concerning posture and locomotion (Robinson, 1972).

The remaining post-cranial bones have not yet been described.

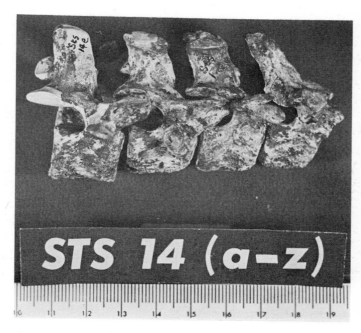

Fig. 79   Four vertebrae from Sterkfontein (Sts 14)
Photographed by courtesy of the Director of the Transvaal Museum, Pretoria

Dimensions   SKULL
Broom, Robinson and Schepers (1950)
Skull 5 Max. Length 146·8   Max. Breadth 99·0
Cranial Index 67·5   Cranial Capacity 482 cc
                                                    485 cc (Holloway,
                                                    1970)
MANDIBLE
Body Thickness at M1 24·0   Body Depth at M1 37·0
Coronoid Height 86·5   Minimum Ramus Breadth *c.* 60·0   **231**

TEETH
Robinson (1956)

Sts 52a		Upper Permanent Teeth (Crown Dimensions)							
		I1	I2	C	PM1	PM2	M1	M2	M3
Left	l	9·5	7·3	9·8	8·7	9·1	12·3	13·2	—
Side	b	8·2	7·0	9·9	12·8	(13·2)	14·0	15·2	—
Right	l	9·3	6·8	9·9	8·6	9·3	12·2	13·3	12·5
Side	b	8·3	7·0	9·7	12·8	13·3	14·1	15·2	14·6

Sts 52b		Lower Permanent Teeth (Crown Dimensions)							
Left	l	—	—	—	—	—	13·4	14·8	13·5
Side	b	—	—	—	—	—	—	13·3	12·9
Right	l	5·9	7·1	9·1	9·0	9·8	13·0	14·4	13·7
Side	b	8·1	8·1	10·0	11·7	11·7	12·9	13·4	12·7

Sts 2		Upper Deciduous Teeth (Crown Dimensions)				
		DI1	DI2	DC	DM1	DM2
Left	l	—	—	—	—	—
Side	b	—	—	—	—	—
Right	l	—	—	—	9·9	11·2
Side	b	—	—	—	9·6	11·3

Sts 24		Lower Deciduous Teeth (Crown Dimensions)				
Left	l	4·2	—	6·4	—	—
Side	b	—	—	5·6	—	—
Right	l	—	—	6·3	8·2	10·7
Side	b	—	—	5·6	6·9	9·0

POST-CRANIAL BONES
Scapula: Glenoid Cavity Max. Length *c*. 33
            Max. Breadth 20
Humerus: Diameter of Head *c*. 40
          Length Between 290 and 310
Innominate Bone: Ant. Sup. Spine–Post. Sup. Spine *c*. 115
           Sacral articulation, Max. Length 29
Femur: Max. Length *c*. 310

Affinities The affinities of the Sterkfontein remains will be discussed with the rest of the australopithecine material.

The Sterkfontein Remains

Originals The Transvaal Museum, Pretoria, Republic of South Africa.

Casts The University Museum, University of Pennsylvania, Philadelphia 4, Pennsylvania, U.S.A. (Proximal end of right humerus, innominate bone, distal ends of right and left femora.)

References BROOM, R. 1936
A new fossil anthropoid skull from South Africa. *Nature 138*, 486-488.
BROOM, R. 1937
The Sterkfontein ape. *Nature 139*, 326.
BROOM, R. and SCHEPERS, G. W. H. 1946
The South African fossil ape-men, the *Australopithecinae. Transv. Mus. Mem. 2*, 1-272.
KERN, H. M. and STRAUS, W. L. 1949
The femur of *Plesianthropus transvaalensis. Am. J. Phys. anthrop. 7*, 53-77.
BROOM, R., ROBINSON, J. T., and SCHEPERS, G. W. H. 1950
Sterkfontein ape-man, *Plesianthropus. Transv. Mus. Mem. 4*, 1-117.
MAYR, E. 1950
Taxonomic categories in fossil hominids. *Cold Spring Harbour Symposia on Quantitative Biology 15*, 109-118.
OAKLEY, K. P. 1954
Dating the australopithecines of Africa. *Am. J. Phys. Anthrop. 12*, 9-23.
ROBINSON, J. T. 1954
The genera and species of the *Australopithecinae. Am. J. Phys. Anthrop. 12*, 181-200.
HOWELL, F. C. 1955
The age of the australopithecines of Southern Africa. *Am. J. Phys. Anthrop. 13*, 635-662.
ROBINSON, J. T. 1956
The dentition of the *Australopithecinae. Transv. Mus. Mem. 9*, 1-179.
BRAIN, C. K. 1957
New evidence for the correlation of the Transvaal ape-man bearing cave deposits. *Proc. Third. Pan-African Cong. Prehist., Livingstone, 1955*, 143-148. Ed. J. D. Clark. London: Chatto and Windus.
EWER, R. F. 1957
Faunal evidence on the dating of the *Australopithecinae. Ibid.*, 135-142.
OAKLEY, K. P. 1957
Dating the australopithecines. *Ibid.*, 155-157.
ROBINSON, J. T., and MASON, R. J. 1957
Occurrence of stone artefacts with *Australopithecus* at Sterkfontein. *Nature 180*, 521-524.

The Sterkfontein Remains

BRAIN, C. K. 1958
The Transvaal ape-man bearing cave deposits. *Transv. Mus. Mem.*
11, 1-125.
ROBINSON, J. T., 1958
The Sterkfontein tool-maker. *The Leech (Johannesburg) 28*, 94-100.
ROBSINSON, J. T., and MASON, R. J. 1962
Australopithecines and artefacts at Sterkfontein. *S. Afr. Arch. Bull.*
17, 87-125.
KURTÉN, B. 1962
The relative ages of the australopithecines of Transvaal and the
pithecanthropines of Java. In *Evolution und Hominisation*, 74-80. Ed.
G. Kurth. Stuttgart: Gustav Fischer Verlag.
COOKE, H. B. S. 1964
Pleistocene mammal faunas of Africa, with particular reference to
South Africa. In *African ecology and human evolution*, 65-116. Eds.
F. C. Howell and F. Boulière. London: Methuen and Co. Ltd.
TOBIAS, P. V., and HUGHES, A. R. 1969
The new Witwatersrand University excavation at Sterkfontein.
S. Afr. Archaeol. Bull. 24, 158-169.
COOKE, H. B. S. 1970
Notes from members: Dalhousie University, Halifax, Canada. *Bull.*
Soc. Vertebra. Paleont. 90, 2.
HOLLOWAY, R. L. 1970
New endocranial values for the australopithecines. *Nature 227*,
199-200.
ROBINSON, J. T. 1972
Early Hominid Posture and Locomotion. Chicago: Chicago University
Press.
DAY, M. H. 1973
Locomotor features of the lower limb in hominids. *Symp. Zool. Soc.*
Lond. 33, 29-51.
PARTRIDGE, T. 1973
Geomorphological dating of cave openings at Makapansgat,
Sterkfontein, Swartkrans and Taung. *Nature, 246*, 75-79.
TOBIAS, P. V. 1973
A new chapter in the history of the Sterkfontein early hominid site.
J. S. Afr. Biol. Soc. 14, 30-44.
BUTZER, K. W. 1974
Curr. Anthrop. Comment. Reply. *Curr. Anthrop. 15*, 413-416.
VRBA, E. S. 1975
Some evidence of chronology and palaeoecology of Sterkfontein,
Swartkrans and Kromdraai from the fossil Bovidae. *Nature 254*,
301-304.
HUGHES, A. R. and TOBIAS, P. V. 1977
A fossil skull probably of the genus *Homo* from Sterkfontein,
Transvaal. *Nature 265*, 310-312.

The Kromdraai Remains

Synonyms *Paranthropus robustus* (Broom, 1938); *Homo transvaalensis*
and other names (Mayr, 1950); *Paranthropus robustus robustus* (Robinson, 1954);
Australopithecus robustus (Oakley, 1954); *Australopithecus
robustus robustus* (Campbell, 1964)
'Near-man'; 'Ape-man'; 'Man-ape'

Site Kromdraai, two miles east of Sterkfontein, nine miles north-
west of Krugersdorp, near Johannesburg, Transvaal, Republic
of South Africa.

Found by (*a*) G. Terblanche, June, 1938; recognized by R. Broom.
Cranial and post-cranial bones.
(*b*) R. Broom, February, 1941. Juvenile mandible.

Geology The bones were found in a block of stony breccia loose on the
surface of the hillside at a point later named Kromdraai B.
Excavation showed that this block was derived from the
filling of a cave, formed in Pre-Cambrian dolomitic limestone, whose roof had completely weathered away. A similar
cave-filling near by, Kromdraai A, has yielded quantities of
mammalian fossil bones but no hominid material. The geology of the site has been investigated in detail (Brain, 1957,
1958).

Associated finds One unquestionable chert artefact has been claimed from
Kromdraai B and four other specimens from the same site
regarded as less convincing stone tools (Brain, 1958). Until
more evidence is available it would be unwise to classify
these implements as a recognizable culture.
Fossil mammalian bones recovered from Kromdraai A
include insectivores (*Proamblysomus antiquus, Elephantulus langi,
Crocidura cf. bicolor, Suncus cf. etruscus*), primates (*Gorgopithecus
major, Parapapio jonesi, Papio angusticeps, Papio robinsoni*), a
number of extinct rodents, several large and small carnivores
(*Canis terblanchei, Herpestes mesotes, Crocuta spelaea, Crocuta
ultra, Felis crassidens, Panthera shawi*), elephant (*Loxodonta
atlantica*), hyraces (*Procavia antiqua, P. transvaalensis*), equids
(*Stylohipparion steytleri, Equus plicatus, Equus helmei*) and a pig
(*Potamochoerops antiquus*) (Cooke, 1964).

Dating The fossil bones recovered from Kromdraai cannot at present   235 be dated by direct chemical or radiometric methods since they were preserved in limestone; assessments of their ages must rely therefore on geological and faunal evidence.

Unfortunately the cave deposit is not clearly stratified, and the artefacts that were found do not form part of an established sequence. Faunal dating also presents problems in South Africa since the Plio-Pleistocene boundary is not clearly defined by a change in the composition of the fauna (Ewer, 1957). Kurtén (1962) attributed the Kromdraai deposit to the Second Glaciation (Mindel or Antepenultimate Glaciation), on the grounds of faunal correlation.

However, Kromdraai is commonly regarded as one of the most recent of the Transvaal australopithecine sites and has been attributed to the Basal Middle Pleistocene (Oakley, 1954, 1964). This conclusion has been supported by a recent study of the bovid remains from the site (Vrba, 1975) which shows Kromdraai B, the hominid fossil-bearing layer, to be younger than Kromdraai A.

Morphology   The fossil hominid bones recovered from Kromdraai B comprise the left half of a cranium including the left maxilla and zygomatic bones, part of the left sphenoid, the left temporal, a fragmentary right maxilla, the right half of the body of a mandible, three isolated molars and four premolars. In addition some post-cranial bones were found including the distal end of a right humerus, the proximal end of a right ulna, and a broken talus. Doubts have been cast on the correctness of the hominid identification of a metacarpal and a number of phalanges recovered from the site (Day and Scheuer, 1973; Day, in press). Later, a juvenile mandible was found containing most of the deciduous teeth and the right first permanent molar.

SKULL

The skull is heavily built, having a relatively large face and a small cranium; the infra-temporal fossa is deep, suggesting marked post-orbital constriction. The brow ridges are absent but it seems likely that they were prominent, rather than exaggerated.

The position of the foramen magnum is well beneath the skull, and the mastoid process is small. The glenoid fossa is   236 broad and shallow, bounded in front by an articular eminence and behind by a modest post-glenoid tubercle. This arrangement suggests a temporo-mandibular mechanism of human character, a concept borne out by the nature of the wear of the teeth. The tympanic bone is broad and flat, forming the posterior wall of the articular fossa.

The maxilla is broad and flat, bearing a single infra-orbital foramen; there is no sign of a maxillo-premaxillary suture on the anterior aspect of the maxilla which is moderately prognathic. The mandible is represented by the anterior two-thirds of the body on the right side; this is very stout and bears the premolar and molar teeth. There is neither pronounced mandibular torus nor chin and the mental foramina are multiple.

PERMANENT TEETH

The Upper Incisors and Canines are lost but the second incisor and canine sockets are small.

The First Upper Premolars are large and have two rounded cusps separated by a fissure; there is a well marked posterior fovea. The socketed specimen has two buccal roots and a lingual root.

The Second Upper Premolars are similar to the first premolars but a little larger; they also have three roots.

The Upper First Molar is irregularly rhomboidal in shape, has four cusps and no trace of a Carabelli complex.

The Upper Second Molars are four-cusped and similar in shape to the first molars.

The Upper Third Molars are basically four-cusped: the arrangement of the cusps is somewhat simplified.

★   ★   ★

The Lower Incisor and Canine Teeth are missing but their sockets in a symphysial fragment are remarkably small.

The Lower First Premolars are large with two main cusps which are low and rounded. The anterior fovea is deep and there is no cingulum.

The Lower Second Premolars are large and bicuspid, perhaps tending to be molariform.

The Lower First Molars are represented by a worn specimen in the type jaw, and an incompletely erupted specimen in the **237** juvenile mandible. The worn specimen has five cusps and a small sixth cusp, whereas the other tooth is small and has no sixth cusp.

The Lower Second Molar is broken but appears to be larger than the first molar; the cusps show evidence of flat wear.

The Lower Third Molar is in the mandible and is well preserved; it has six cusps and is larger than the other two molars.

DECIDUOUS TEETH

These teeth are known from the mandible of a juvenile whose dental age has been estimated at about three to four years.

The First Lower Deciduous Incisor is absent.

The Second Lower Deciduous Incisor is small and very like the corresponding human tooth.

The Lower Deciduous Canine has a very small crown with a moderately well defined cingulum, and mesial and distal cusplets.

The First Deciduous Molars are both present and unworn. They appear to be remarkably human in their shape and cusp pattern.

The Second Deciduous Molar is rather elongated mesio-distally, has five main cusps and a rudimentary sixth cusp.

POST-CRANIAL BONES

The Lower Extremity of the Right Humerus is well preserved and remarkably human in its general shape. The capitulum is rounded but set a little farther back than is usual in modern man. The medial epicondyle is rather pointed but is marked for the attachment of the flexor muscles of the forearm. The lateral epicondyle is similarly marked for the extensor muscles. According to Le Gros Clark (1947) the humerus has none of the distinctive features found in the recent anthropoid apes; however, this view is not shared by Straus (1948) who suggests that this humerus is 'no more hominid than anthropoid' and believes that its principal affinities are with man *and* chimpanzee. The mixed features of this bone have been confirmed by Patterson and Howells (1967) and McHenry and Corruccini (1974).

The Ulnar Fragment was found near the end of the humerus and almost certainly belongs with it; it resembles that of modern man. The olecranon process is small, suggesting that the range of elbow extension was full.

Fig. 80   A right talus (TM 1517) from Kromdraai
Superior view
Photographed by courtesy of the Director of the Transvaal Museum, Pretoria

The Talus lacks the lower part of the body and head. It is a small bone with a narrow superior articular surface but a broad head and short neck. The horizontal angle of the neck is high by comparison with modern man.

Dimensions   SKULL
Broom and Schepers (1946)
Cranial Capacity 650 cc (estimated)
MANDIBLE
Body Thickness (PM2) 23·4

The Kromdraai Remains

TEETH
Robinson (1956)

		Upper Permanent Teeth (Crown Dimensions)						
		I₁	I₂	C	PM₁	PM₂	M₁	M₂	M₃

		I1	I2	C	PM1	PM2	M1	M2	M3
Left	l	—	—	—	10·0	10·3	13·7	13·8	14·4
side	b	—	—	—	13·7	15·2	14·6	15·9	16·2
Right	l	—	—	—	10·2	—	—	13·8	14·2
side	b	—	—	—	—	—	—	15·9	16·1

		Lower Permanent Teeth (Crown Dimensions)						

		I1	I2	C	PM1	PM2	M1	M2	M3
Left	l	—	—	—	10·0	11·1	—	—	—
side	b	—	—	—	12·2	13·0	—	—	—
Right	l	—	—	—	—	11·0	14·4	—	16·4
side	b	—	—	—	—	13·1	13·0	—	14·0

		Lower Deciduous Teeth (Crown Dimensions)					

		DI1	DI2	DC	DM1	DM2	M1 (Perm.)
Left	l	—	—	—	9·7	12·5*	—
side	b	—	—	—	8·1	10·0	—
Right	l	—	4·6	5·2	9·7	11·6	12·7
side	b	—	3·7	4·9	8·1	9·7	11·5

* Damaged

POST-CRANIAL BONES
Broom and Schepers (1946)
Humerus: Bicondylar Width 54·0
           Max. Width of Articular Surface 40·0
Metacarpal: Length approx. 70·0
             Width of Head 12·0
Proximal Phalanx (? Left 2nd finger): Length 45·0
Distal Phalanx (? 2nd or 3rd toe): Length 12·5
                              Width of proximal end 8·3
Talus (taken from a cast)
  Length 34·5
  Breadth 28·5
  Length/Breadth Index 82·6
  Horizontal Angle of Neck 30°
  Torsion of Neck approx. 30°

Humerus (Straus, 1948; from a cast)
Bicondylar Width 54·0
Width of the Trochlear 20·0
Depth of the Trochlear 23·0
Width of the Capitulum 16·0
Max. Width of Articular Surface 40·0

Affinities The affinities of the Kromdraai remains will be discussed with the rest of the australopithecine material.

Originals The Transvaal Museum, Pretoria, Republic of South Africa.

Casts The University Museum, University of Pennsylvania, Philadelphia 4, Pennsylvania, U.S.A.

References BROOM, R. 1938
The Pleistocene anthropoid apes of South Africa. *Nature 142*, 377 - 379.
BROOM, R. and SCHEPERS, G. W. H. 1946
The South African fossil ape-men, the *Australopithecinae. Transv. Mus. Mem. 2*, 1-272.
CLARK, W. E. LE GROS 1947
Observations on the anatomy of the fossil *Australopithecinae.]. Anat. (Lond.) 81*, 300-333.
STRAUS, W. L. JNR. 1948
The humerus of *Paranthropus robustus. Am. J. Phys. Anthrop. 6*, 285-311.
MAYR, E. 1950
Taxonomic categories in fossil hominids. *Cold Spring Harbour Symposia on Quantitative Biology 15*, 109-118.
OAKLEY, K. P. 1954
Dating the australopithecines of Africa. *Am. J. Phys. Anthrop. 12*, 9-23.
ROBINSON, J. T. 1954
The genera and species of the *Australopithecinae. Am. J. Phys. Anthrop. 12*, 181-200.
ROBINSON, J. T. 1956
The dentition of the *Australopithecinae. Transv. Mus. Mem. 9*, 1-179.
BRAIN, C. K. 1957
New evidence for the correlation of the Transvaal ape-man bearing cave deposits. *Proc. Third Pan-African Cong. Prehist., Livingstone, 1955*, 143-148. Ed. J. D. Clark. London: Chatto and Windus.
EWER, R. F. 1957
Faunal evidence on the dating of the *Australopithecinae. Ibid.*, 135-142.

The Kromdraai Remains

BRAIN, C. K. 1958
The Transvaal ape-man bearing cave deposits. *Transv. Mus. Mem.* *11*, 1-125.

KURTÉN, B. 1962
The relative ages of the australopithecines of Transvaal and the pithecanthropines of Java. In *Evolution und Hominisation*, 74-80. Ed. G. Kurth. Stuttgart: Gustav Fischer Verlag.

OAKLEY, K. P. 1964
Frameworks for dating fossil man, 291. London: Weidenfeld and Nicolson.

CAMPBELL, B. 1964
Quantitative taxonomy and human evolution. In *Classification and human evolution*, 50-74. Ed. S. L. Washburn. London: Methuen and Co. Ltd.

COOKE, H. B. S. 1964
Pleistocene mammal faunas of Africa, with particular reference to South Africa. In *African ecology and human evolution*, 65-116. Eds. F. C. Howell and F. Boulière. London: Methuen and Co. Ltd.

PATTERSON, B., and HOWELLS, W. W. 1967
Hominid humeral fragment from Early Pleistocene of northwestern Kenya. *Science*, 156, 64-66.

DAY, M. H. and SCHEUER, J. L. 1973
SKW 14147: a new hominid metacarpal from Swartkrans. *J. Hum. Evol.* 2, 429-438.

MCHENRY, H. and CORRUCCINI, R. S. 1974
Distal humerus in hominid evolution. *Folia Primatologia*, 227-244.

VRBA, E. S. 1975
Some evidence of chronology and palaeoecology of Sterkfontein, Swartkrans and Kromdraai from the fossil Bovidae. *Nature 254*, 301-304.

DAY, M. H.
Functional interpretations of the morphology of postcranial remains of early African hominids. Ed. C. Jolly. *Wenner Gren Conference, New York, 1974* (in press).

The Swartkrans Remains

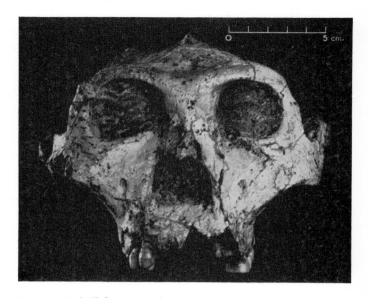

Fig. 81   A skull from Swartkrans, SK 48
Frontal view
Courtesy of Professor J. T. Robinson

Synonyms and other names   1. *Paranthropus crassidens* (Broom, 1949); *Homo transvaalensis* (Mayr, 1950); *Australopithecus (Paranthropus) crassidens* (Oakley, 1954); *Paranthropus robustus crassidens* (Robinson, 1954); *Australopithecus robustus crassidens* (Campbell, 1964); *Paranthropus robustus* (Brain, 1970); 'Ape-man'; 'Man-ape'; 'Near-man'

2. *Telanthropus capensis* (Broom and Robinson, 1949); *Pithecanthropus capensis* (Simonetta, 1957); *Homo erectus* (Robinson, 1961); *Australopithecus capensis* (Oakley, 1964); *Homo sp. indet.* (Clarke, Clark Howell and Brain, 1970).

3. Hominidae, genus and species indet. (Day and Scheuer, 1973).

Site   Swartkrans, six miles north-west of Krugersdorp, near Johannesburg, Republic of South Africa.   243

The Swartkrans Remains

Found by   R. Broom, November, 1948–1951; J. T. Robinson, 1948–1952; J. T. Robinson, April 1949; C. K. Brain, 1966, 1967 and 1968 *et seq.*

Geology   The Swartkrans site is situated near to the Sterkfontein and Kromdraai excavations and consists of the remains of a cavern in Pre-Cambrian dolomitic limestone. The roof of the outer part of the cave has been removed by erosion but the inner cave is still protected by a thick layer of dolomite. The cavern appears to have originated from subsidence of the deposit into an underlying solution cavity.

The cave-filling is made up of two types of breccia, pink (primary) and brown (secondary), above a layer of dripstone. The younger brown breccia of the inner cave overlies the older pink breccia of the outer cave unconformably. The pink breccia contained nearly all of the hominid fossils and is unstratified, whereas the brown breccia contained comparatively little fossil bone and is well stratified. The geology of the site has been investigated in detail (Brain, 1958), and reinterpreted more recently (Brain, 1970).

Associated finds   At first several quartzite artefacts were recovered from the dumps of breccia which surrounded the excavation. They appeared to be of Oldowan type but too few were recovered to allow a definite cultural assessment.

Recently a new assemblage of stone tools has been recovered associated with hominid remains (Brain, 1970). The new assemblage has been studied and the following tool types identified; side and end choppers, bifaces, discoids, subspheroids, scrapers and picks. All are in sharp condition. By comparison with Oldowan tools from Olduvai they are relatively large, but otherwise they show a number of similarities (Leakey, 1970).

The fossil mammalian bones that were found include those of insectivores (*Chlorotalpa spelea, Elephantulus langi, Elephantulus* cf. *brachyrhynchus, Suncus* cf. *etruscus*), primates (*Simopithecus danieli, Parapapio jonesi, Papio robinsoni, Dinopithecus ingens, Cercopithecoides williamsi*), rodents (*Dasymys bolti, Palaeotomys gracilis, Cryptomys robertsi*), several carnivores (*Canis mesomelas pappos, Cynictis penicillata, Lycyaena silbergi, Lycyaena nitidula, Leecyaena forfex, Crocuta crocuta angella, Crocuta* 244 *venustula*, *Hyaena brunnea*, *H. brunnea dispar*, *Panthera* aff. *leo*, *P. pardus incurva*, *Megantereon eurynodon*), hyraces (*Procavia antiqua*, *P. transvaalensis*) and two artiodactyls ('*Tapinochoerus' meadowsi*, *Potamochoerops antiquus*) (Cooke, 1964).

Dating The nature of the deposit at Swartkrans precludes chemical or radiometric dating with the techniques available at present. The dating problem has been discussed by a number of authors (Howell, 1955; Ewer, 1957; Oakley, 1954, 1957; Brain, 1958). Kurtén (1962) has suggested recently that all the Transvaal australopithecine sites are post-Villafranchian and that Swartkrans should be attributed to the Second Glaciation (Mindel Glaciation or Antepenultimate Glaciation) with Kromdraai, on the grounds of faunal correlation. This later date remains to be established since the balance of faunal and geological evidence seemed to indicate a Basal Middle Pleistocene age for both of these sites (Oakley, 1954, 1964). Recently it has been suggested that the opening of the Swartkrans cave should be dated at 2·57 million years B.P. on the basis of a geomorphological method (Partridge 1973). The method has been criticized and is not universally accepted as a valid technique (Butzer, 1974). Vrba (1975) has suggested a date of approximately 1·5 million years B.P. for this site on the grounds of bovid faunal associations. No radiometric, chemical, palaeomagnetic or biochemical methods have proved usable at Swartkrans to assist with the dating problem.

Morphology The remains from Swartkrans include:
1. An almost complete cranium, the left half of a cranium with both maxillae, a juvenile cranium, an adolescent cranium and several other crania which have been crushed during fossilization. A number of specimens of maxillae, with most of the upper dentition, were recovered as well as three adult mandibles. Several other mandibles were found, both adult and juvenile, and about 100 isolated teeth. The post-cranial bones include the crushed lower end of a humerus, a left thumb metacarpal, an incomplete innominate bone and the proximal ends of two right femora. An infant mandible containing deciduous molars (SK 3978), a natural endocast of a hominid skull (SK 1585), a thoracic vertebra (T_{12}) and a lumbar vertebra (L_5) (SK 3981). In addition to these important 245

The Swartkrans Remains

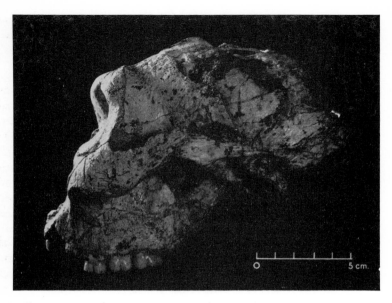

Fig. 82   A skull from Swartkrans, SK 48
Left lateral view
Courtesy of Professor J. T. Robinson

specimens there have been recovered a number of mandibular
and cranial fragments as well as a dozen isolated teeth (Brain,
1970).
2. A mandible (SK 15), a mandibular fragment (SK 45)
containing two molar teeth, a maxillary fragment (SK 80),
an isolated lower third premolar (SK 18a), the proximal end
of a radius (SK 18b) and the distal end of a fourth left metacarpal (SK 85). Following a recent examination of the
Swartkrans remains (Clarke, Clark Howell and Brain, 1970),
one of these authors (R. J. Clarke) was able to associate parts
of a skull (SK 847) and much of a left temporal bone
(SK 846b), both formerly attributed to *A. robustus*, with the
previously known maxillary fragment (SK 80). The specimen
comprises most of the left side of the face, the infratemporal
fossa and the left temporal region. The whole unit is now
termed SK 847.   246

3. Subsequently an adult left fifth metacarpal (SKW 14147) was recovered from a block of breccia (Day and Scheuer, 1973).

I. SKULL

The best cranium (SK 48) obtained from Swartkrans is probably that of a female; it is somewhat crushed and was broken during excavation. The supra-orbital ridge is well marked, particularly in the midline, showing a prominent glabella; the forehead is concave and restricted by the union of the temporal lines forming a sagittal crest. The brain-case is small in relation to the size of the facial skeleton, and constricted behind the orbits. Posteriorly the occipital region is damaged but the occipital crest is well defined. The glenoid fossa is deep and bounded posteriorly by the tympanic bone.

The maxillae are robust and flat, separated by a broad nasal opening, but the premaxillary region is shortened giving the dental arcade a 'squared-off' appearance. The palate is large and bounded by stout alveolar processes of bone bearing three molars on the left side and the first premolar and canine on the right.

MANDIBLE

The best-preserved mandible (SK 23) is a fine specimen, virtually complete, with an almost perfect adult dentition. The only damage it has sustained is some crushing of the body which has resulted in fracture of the symphysial region and narrowing of the dental arcade. This specimen is probably female since two other jaws from Swartkrans are similar in their general features but are even more massive.

Fig. 83   A mandible from Swartkrans, SK 23
Courtesy of Professor J. T. Robinson

The body of the female mandible is stout; the ramus is very tall and buttressed internally by paired ridges running to the coronoid and condyloid processes from the *crista pharyngea,* a crest which runs upwards from the lingual side of the third molar. The symphysial region, although broken, shows that there is neither chin nor true simian shelf, but there is a genioglossal fossa internally in the midline below the flattened alveolar plane. The internal mandibular torus is of moderate size; externally the anterior border of the coronoid process runs down on to the body as a ridge which passes below the molar teeth and fades away beneath the single mental foramen.

PERMANENT TEETH

About 300 permanent teeth are known from Swartkrans.

The Upper Incisors are relatively small and shovel-shaped with raised marginal ridges. The lateral incisors tend to be smaller than the central incisors.

The Upper Canines are small-crowned and stout-rooted; they are symmetrical and have blunt points. The lingual faces of these teeth have characteristic grooves.

The Upper First Premolars are bicuspid and asymmetrical in occlusal view; the primary fissure is usually deeply cut.

The Upper Second Premolars resemble the first premolars in cusp pattern but usually have a distinct talon.

The Upper First Molars are characteristically trapezoidal in shape and the crown pattern is simple and four-cusped.

The Upper Second Molars are also skewed in occlusal view and several examples have some of the Carabelli complex.

The Upper Third Molars resemble the second molars in some respects but tend to be even more irregular in shape.

★   ★   ★

The Lower Incisors are relatively simple having horizontal cutting edges; the lateral specimens are slightly shovel-shaped.

The Lower Canines are asymmetrical having a distal cusplet formed by the remnants of the cingulum.

The Lower First Premolars are robust, bicuspid and asymmetrical. The buccal cusp is the larger and it is commonly joined to the lingual cusp by a ridge which separates the anterior and posterior foveae.   248

The Lower Second Premolar is larger than the first and its crown is partially molarized.

The Lower First Molars are typically hominid in shape; five main cusps are present as well as a small sixth cusp. The fissure pattern is normally dryopithecine with some tendency towards the development of the + pattern typical of modern man. There is some wrinkling of the occlusal enamel.

The Lower Second Molars are similar in shape to the first molars but have four main cusps and a small fifth cusp. The fissure pattern is not dryopithecine and there is some secondary enamel wrinkling.

The Lower Third Molars have complex occlusal cusp patterns and crenelated fissure arrangements. The tooth crowns tend to be of irregular shape.

DECIDUOUS TEETH

The Upper Incisors and Canines are not represented well enough for description.

The Upper First Molar is known from a single specimen, a very worn isolated tooth. The crown is asymmetrical and has four cusps.

The Upper Second Molars, though smaller than the first permanent molars, are very like them in the details of their structure and also have four main cusps.

★   ★   ★

The Lower Incisors are too badly damaged for description.

The Lower Canines have asymmetrical crowns with well marked distal cusplets derived from the cingulum.

The Lower First Molars are well developed and have five cusps separated by the basically dryopithecine fissure system.

The Lower Second Molars are larger than the first and fully molariform. The cusp pattern agrees closely with that of the first molar.

POST-CRANIAL BONES

The Left Thumb Metacarpal (SK 84) is robust, curved and strongly impressed by muscular markings. The distal articular surface is asymmetrical and has a beak-like process separating two sesamoid grooves. The proximal articulation is saddle-shaped. *The Innominate Bone* (SK 50) belongs to the right side and lacks **249**

The Swartkrans Remains

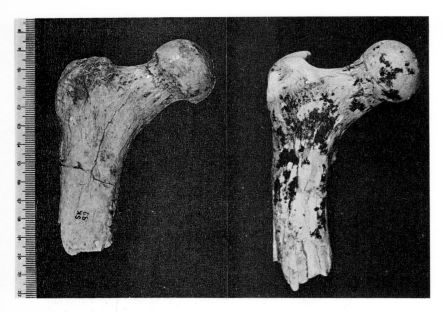

Fig. 84   The upper ends of two femora (*left:* SK 97 *right:* SK 82) from Swartkrans
Anterior views
Photographed by courtesy of the Director of the Transvaal Museum, Pretoria

the iliac crest, the sacral articulation and most of its pubic portion. The acetabulum is complete but crushing has reduced its
anteroposterior diameter. The anterior superior iliac spine is
prominent and reaches forwards and laterally; although the
iliac crest is missing it seems likely that it is less sharply
curved than that of man. The ischial segment is long and the
ischial tuberosity well separated from the acetabular margin.
The distortion and generally poor state of preservation of
this fossil hip bone precludes the use of many measurements
for comparative purposes.

The Upper Ends of Two Femora (SK 82 and SK 97), both from
the right side, are similar in their principal features. The head
is small and rounded, bearing a sub-central fovea. The neck,
broad at the base, narrows to the articular surface, has an
horizontal upper border and a pronounced obturator externus   250

groove indicating clearly that the hip was capable of hyper-extension (Day, 1969). There is a weak trochanteric crest but the great trochanter is hollowed by a deep pit towards its apex, whilst the lesser trochanter is directed posteriorly and medially. The small size of the head by comparison with the thickness of the shaft and the development of the trochanters gives the bone a disproportionate appearance.

Recently two vertebrae have been recovered from the Swartkrans cave site (Brain, 1970). It appears that they are derived from a block of breccia similar to that from which the early hominid materials are derived. They have been

Fig. 85   The Swartkrans innominate bone, SK 50
Courtesy of Professor J. T. Robinson

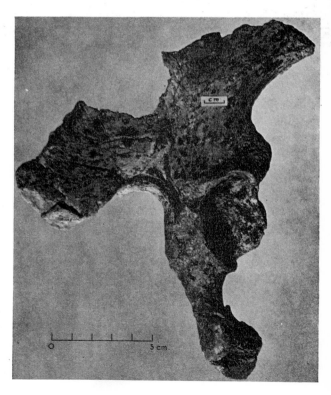

identified as being a last thoracic vertebra (SK 3981a) and a last lumbar vertebra (SK 3981b).

They both are large and have general resemblances to the australopithecine vertebrae from Sterkfontein (Sts 14); they differ from those of modern man in several respects. In view of this, they have been attributed to *Paranthropus robustus* (Robinson, 1970). In functional terms the most important feature of the last lumbar vertebra may be the disparity between the anterior and posterior depths of the body; this wedging suggests a well developed sacral promontory and marked lumbar lordosis. In turn this supports the view that *Paranthropus robustus* (≡ *A. robustus*) had habitual upright posture.

Fig. 86   A mandible from Swartkrans, SK 15
(Telanthropus I)
Occlusal view
Courtesy of Professor J. T. Robinson

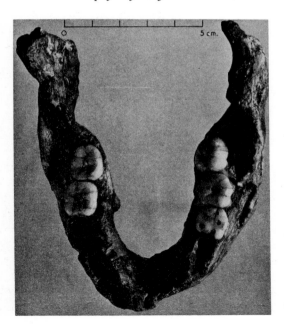

2. MANDIBLE

In the Swartkrans cave two mandibles were found whose morphology is very different from that of those described in the preceding section. The body of the better specimen (SK 15) is complete and contains five molar teeth, but the ascending rami are broken; the whole jaw is somewhat crushed. The symphysis is almost intact and has neither chin nor simian shelf; internally it is reinforced by a transverse torus below which there is a shallow genioglossal fossa. The body is robust, particularly in the region of the third molar, and the mental foramen on the right side is single.

The second specimen (SK 45) consists of a fragment of the right side of the body of another mandible bearing two molar teeth. An isolated left premolar tooth (SK 18a) was found close to the better mandible.

TEETH

The Premolar is considerably worn but the crown is small and hominid in its general features.

The Left First Molar is intact but worn; however, it is not distinguishable from a modern human lower first molar.

The Second Molars are both worn but it is possible to discern a dryopithecine fissure pattern with six cusps. There seems to be no trace of enamel wrinkling.

The Third Molars are intact and little worn, and resemble the second molars. The dryopithecine fissure pattern (Y5) is tending towards the + pattern of modern hominids.

THE NEW SKULL (SK 847)

This skull, constructed from materials that were formerly attributed to both *A. robustus* and '*Telanthropus*', shows a number of distinctive features. The supraorbital torus is pronounced and quite thick while the postorbital constriction is not marked. The nasal bones are prominent on the face and differ strongly from those of both gracile and robust australopithecines known previously. The position of the articular fossa for the mandibular condyle with relation to the occlusal plane is such that the ramus of the mandible must have been squat and quite different in morphology from the tall *A. robustus* mandible (SK 23) known from the same site.   253

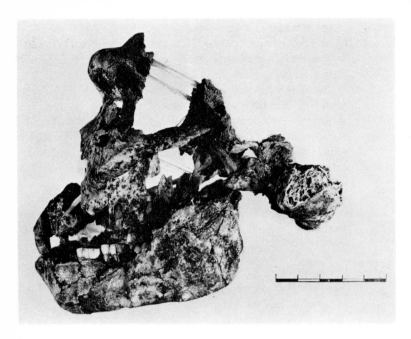

Fig. 87   The Swartkrans (SK 847) composite cranium articulated with the
SK 15 mandible
Courtesy of R. J. Clarke

In fact the '*Telanthropus*' I mandible (SK 15) 'fits' SK 847 very
well although it is dentally not from the same specimen. The
other mandibular fragment (SK 45), however, could well
represent the same individual.

POST-CRANIAL BONES
The proximal end of a right *Radius* (SK 18b) was recovered
with the better mandible. It is said to be indistiguishable from
that of modern man although it has never been fully described.
The distal end of a fourth left *Metacarpal* (SK 85) was found at
Swartkrans. It is not particularly robust, nor does it have any
of the specializations associated with pongid metacarpals.
Apart from its small size it is similar to the corresponding
bone of modern man.

254

Dimensions I. SKULLS
None of the crania is sufficiently well preserved to allow reliable measurements of their principal dimensions or cranial capacities.
Natural endocast (SK 1585)—volume 530 cc
Holloway (in Brain, 1970)

MANDIBLE (SK 23)
Robinson (1953)
Max. Length 127·0
Ramus Height 91 (at coronoid); 84 (at condyle)
Ramus Width 52 and 55 (mid-point)
Body Depth (PM2) 39·7
Body Depth (C) 40·0
Body Depth (M2) 35·7
Body Depth (M3) 33·0
Body Width (M2) 25·0

TEETH
Robinson (1956)

		Upper Permanent Teeth (Crown Dimensions)						
		I1	I2	C	PM1	PM2	M1	M2	M3
Average	l	9·4	7·2	8·7	9·9	10·6	13·8	14·7	15·1
figures	b	7·6	6·8	9·5	14·2	15·4	14·5	15·9	16·9

SK 23		Lower Permanent Teeth (Crown Dimensions)						
Left	l	5·6	6·7	7·8	9·4	10·5	14·8	15·2	16·0
side	b	5·9	6·7	7·8	11·4	14·1	14·7	14·8	13·0
Right	l	5·6	6·6	8·1	9·6	10·7	14·7	15·0	17·3
side	b	6·3	6·7	8·0	11·5	13·5	14·6	14·8	14·1

		Upper Deciduous Teeth (Crown Dimensions)				
		DI1	DI2	DC	DM1	DM2
Left	l	—	—	—	—	10·2
side	b	—	—	—	—	12·0
Right	l	—	—	—	8·7	—
side	b	—	—	—	9·8	—
Specimens					SK 91	SK 90

SK 61		Lower Deciduous Teeth (Crown Dimensions)				
		DI1	DI2	DC	DM1	DM2
Left	l	—	—	5·7	10·6	13·3
side	b	—	—	5·3	9·5	11·9
Right	l	3·8	4·9	5·9	11·1	13·4
side	b	—	4·6	5·2	9·5	12·0

Brain (1970)

SK 3978		Lower Deciduous Teeth (Crown Dimensions)				
		DI_1	DI_2	DC	DM_1	DM_2
Left	l	—	—	—	10·2	13·0
side	b	—	—	—	7·8	10·5
Right	l	—	—	—	9·9	12·9
side	b	—	—	—	8·2	10·7

POST-CRANIAL BONES
The Thumb Metacarpal (SK 84) (Napier, 1959)
Max. Length 35·0   Mid-shaft Breadth 7·5 (A.P.),   9·5 (Transverse)   Robusticity Index 24·3
Femur (SK 82)
Length of neck 63·0 (from a cast)
Mean Thickness of shaft 27·5 (below lesser trochanter) (from a cast)
Mean Diameter of head 33·0 (from a cast)

SK 3981a (T_{12})   (Robinson, 1970)

No dimensions available.

SK 3981b (L_5)
Width of neural canal—21·4

Body   Anterior depth—21·3
         Posterior depth—19·0

2. SKULL
Clarke, Clark Howell and Brain (1970)

SK 847

Parietal/sagittal chord (damaged) 82·5
Max. thickness (22 mm. anterior to lambda) 7·5   256

Max. thickness (posterior part of temporal line) 8·5
Distance of temporal line from sagittal suture at lambda 37·0
Max. distance of temporal line from sagittal suture 39·0

MANDIBLES
Robinson (1953)
Telanthropus I (Restored) (SK 15)
Max. Length 109·0
Body Height (C) 31·5
Body Height (M3) 25·0
Body Width (M2) 22·5
Ramus Height (coronoid) 59·0
Ramus Height (condyle). 55·0
Bicondylar Width 114·0

TEETH
Broom and Robinson (1952)

Lower Permanent Teeth (Crown Dimensions)				
Telanthropus I (SK 15)	*PM1*	*M1*	*M2*	*M3*
Left l	8·6	11·9	—	14·3
side b	10·3	11·9	—	12·4
Right l	—	—	13·6	13·9
side b	—	—	13·1	12·3

POST-CRANIAL BONES
Fourth Left Metacarpal (SK 85) (Napier, 1959)
Reconstructed Length 50·7
No dimensions are available for the upper end of the radius.

3. SKW 14147 (Day and Scheuer, 1973)

Left fifth metacarpal
Length 47·4
Mean midshaft thickness 7·1

Affinities The affinities of the Swartkrans remains will be discussed with the rest of the australopithecine material.
Originals The Transvaal Museum, Pretoria, Republic of South Africa.
Casts The University Museum, University of Pennsylvania, Philadelphia 4, Pennsylvania, U.S.A. 1. Mandible, skull, immature skull, innominate, femora, palate, maxillae. 2. Mandible. 3. Not available at present.

References   BROOM, R. 1949
Another new type of fossil ape-man. *Nature 163*, 57.
BROOM, R., and ROBINSON, J. T. 1949
A new type of fossil man. *Nature 164*, 322-323.
BROOM, R., and ROBINSON, J. T. 1950
Man contemporaneous with the Swartkrans ape-man. *Am. J. Phys. Anthrop. 8*, 151-156.
MAYR, E. 1950
Taxonomic categories in fossil hominids. *Cold Spring Harbour Symposia on Quantitative Biology 15*, 109-118.
BROOM, R., and ROBINSON, J. T. 1952
Swartkrans ape-man. *Paranthropus crassidens. Transv. Mus. Mem. 6*, 1-124.
ROBINSON, J. T. 1953
Telanthropus and its phylogenetic significance. *Am. J. Phys. Anthrop. 11*, 445-501.
OAKLEY, K. P. 1954
Dating the australopithecines of Africa. *Am. J. Phys. Anthrop. 12*, 9-23.
ROBINSON, J. T. 1954
The genera and species of the *Australopithecinae. Am. J. Phys. Anthrop. 12*, 181-200.
CLARK, W. E. LE GROS 1955
The *os innominatum* of the recent *Ponginae* with special reference to that of the *Australopithecinae. Am. J. Phys. Anthrop. 13*, 19-27.
DART, R. A. 1955
Australopithecus prometheus and *Telanthropus capensis. Am. J. Phys. Anthrop. 13*, 67-96.
HOWELL, F. C. 1955
The age of the australopithecines of Southern Africa. *Am. J. Phys. Anthrop. 13*, 635-662.
ROBINSON, J. T. 1956
The dentition of the *Australopithecinae. Transv. Mus. Mem. 9*, 1-179.
BRAIN, C. K. 1957
New evidence for the correlation of the Transvaal ape-man bearing cave deposits. *Proc. Third Pan-African Cong. Prehist., Livingstone, 1955.* Ed. J. D. Clark, pp. 143-148. London: Chatto and Windus.
EWER, R. F. 1957
Faunal evidence for the dating of the *Australopithecinae. Ibid.*, pp. 135-142.
OAKLEY, K. P. 1957
Dating the australopithecines. *Ibid.*, pp. 155-157.
SIMONETTA, A. 1957
Catalogo e sinominia annotata degli ominoidi fossili ed attuali (1758-1955). *Atti. Soc. tosc. Sci. Nat. 64*, 53-112.
BRAIN, C. K. 1958
The Transvaal ape-man bearing cave deposits. *Transv. Mus. Mem. 11*, 1-125.

The Swartkrans Remains

NAPIER, J. R. 1959
Fossil metacarpals from Swartkrans. Fossil Mammals of Africa, No. 17. London: British Museum (Nat. Hist.)

ROBINSON, J. T. 1961
The australopithecines and their bearing on the origin of man and of stone tool-making. *S. Afr. J. Sci.* *57*, 3-13.

KURTÉN, B. 1962
The relative ages of the australopithecines of Transvaal and the pithecanthropines of Java. In *Evolution und Hominisation*, 74-80. Ed. G. Kurth. Stuttgart: Gustav Fischer Verlag.

CAMPBELL, B. 1964
Quantitative taxonomy and human evolution. In *Classification and human evolution*, 50-74. Ed. S. L. Washburn. London: Methuen and Co. Ltd.

COOKE, H. B. S. 1964
Pleistocene mammal faunas of Africa, with particular reference to South Africa. In *African ecology and human evolution*, 65-116. Eds. F. C. Howell and F. Bourlière. London: Methuen and Co. Ltd.

NAPIER, J. R. 1964
The evolution of bipedal walking in the hominids. *Arch. Biol.* (*Liège*) *75*, Supp., 673-708.

OAKLEY, K. P. 1964
Frameworks for dating fossil man, 291. London: Weidenfeld and Nicolson.

DAY, M. H. 1962
Femoral fragment of a robust australopithecine from Olduvai Gorge, Tanzania. *Nature 221*, 230-233.

WOLPOFF, M. H. 1968
'Telanthropus' and the single species hypothesis. *Am. Anthrop. 70*, 477-493.

DAY, M. H. 1969
A robust australopithecine femoral fragment from Olduvai Gorge, Tanzania (Hominid 20). *Nature 221*, 230-233.

BRAIN, C. K. 1970
New finds at the Swartkrans site. *Nature 225*, 1112-1119.

ROBINSON, J. T. 1970
Two new early hominid vertebrae from Swartkrans. *Nature 225*, 1217-1219.

CLARKE, R. J., HOWELL, F. CLARK and BRAIN, C. K. 1970
More evidence of an advanced hominid at Swartkrans. *Nature 225*, 1219-1222.

LEAKEY, M. D. 1970
Stone artefacts from Swartkrans. *Nature 225*, 1222-1225.

DAY, M. H. and SCHEUER, J. L. 1973
SKW 14147: a new metacarpal from Swartkrans. *J. Hum. Evol. 2*, 429-438.

The Makapansgat Remains

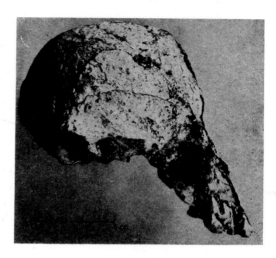

Fig. 88   An incomplete cranium (MLD 37 and
MLD 38) from Makapansgat; most of the
facial parts have weathered away
*Courtesy of Professor P. V. Tobias, photo-
graphed by A. R. Hughes*

Synonyms   *Australopithecus prometheus* (Dart, 1948a); *Homo transvaalensis*
and other names   (Mayr, 1950); *Australopithecus africanus transvaalensis* (Robinson, 1954)
'Ape-man'; 'Man-ape'; 'Near-man'

Site   Makapansgat Limeworks Dump, Makapansgat Farm, 13
miles north-east of Potgieterus, Central Transvaal, Republic of
South Africa.

Found by   Professor R. A. Dart, the staff and students of the Department
of Anatomy, University of the Witwatersrand, Johannesburg,
and other helpers who included J. Kitching, S. Kitching,
B. Kitching, A. R. Hughes, E. L. Boné, R. S. Cunliff and
B. Maguire. The finds were made between September 1947
and 1962.

Geology   The limeworks site is situated on the south side of the Makapan
valley and consists of a large infilled subsidence cavern which   260

has formed in the Transvaal dolomitic limestone of the region. The major part of the cave roof has eroded away and exposed the consolidated cave breccia on the surface. The bedrock of the cave is covered by a considerable layer of pure dripstone indicating that the cave did not communicate with the surface at the time of its accumulation. When the cave opened to the surface, breccia was formed in two principal phases. Most of the hominid material was found in the lower part of the Phase I breccia. The geology of the site has been studied in detail (King, 1951; Brain, 1958).

Associated finds   Numerous utilized bones, teeth and horns have been claimed from the vast accumulation of fossil material from these deposits (Dart, 1957a). Opinion is divided as to whether their appearance is artefactual or natural. A recent review of the problem of the osteodontokeratic (bone, tooth and horn) culture of the australopithecines (Wolberg, 1970) has given some support to Dart's contentions as to the nature of the bone accumulations and the uses to which they may have been put. The debate continued later (Dart, 1971; Wolberg, 1971) and has been expanded to take in the topic of scavenging as an early hominid activity (Read, Martin and Read, 1975).

The fossil mammalian fauna recovered from this site includes a bat (*Rhinolophus cf. capensis*), primates (*Simopithecus darti, Parapapio jonesi, P. broomi, P. whitei, Cercopithecoides williamsi*), numerous rodents, some carnivores (*Cynictis penicillata brachydon, Crocuta cf. brevirostris, Hyaena makapani, Therailurus barlowi, Megantereon sp. nov.*), an equid (*Equus helmei*), perissodactyla (*cf. Ceratotherium simum, cf. Diceros bicornis*) and numerous artiodactyls including antelopes, gazelles, pigs and giraffes (Cooke, 1964).

Dating   As with the other australopithecine sites in the Transvaal, accurate dating has proved difficult because of the nature of the deposits; the problem has been discussed by several authors (Howell, 1955; Ewer, 1957; Oakley, 1954, 1957; Brain, 1958). Later the Makapansgat site was attributed to the First Inter-glacial (Günz–Mindel or Antepenultimate Interglacial) on the grounds of faunal correlation (Kurtén, 1962). A reappraisal of the fauna (Cooke, 1970), in the light of East African evidence, has led to the suggestion that this site should be dated to a   261

The Makapansgat Remains

period 2·5–3·00 million years B.P. Partridge (1973) using a geomorphological method, dates the opening of the Makapansgat cave to an even earlier date, approximately 3·67 million years B.P. Butzer (1974) following a reappraisal of the Taung site concludes that Makapansgat clearly antedates the Taung remains, while not accepting the validity of Partridge's geomorphological dating method.

Morphology The hominid remains recovered from the excavation include a cranium, parts of other crania, a cranio-facial fragment, parts of a number of mandibles, teeth, parts of maxillae, two left ilia, a right ischium and fragments of humerus, radius, clavicle and femur.

SKULL

The best cranium (MLD 37 and MLD 38) (Dart, 1962a) was found split in half in a divided block of Upper Phase I pink

Fig. 89   The first Australopithecine cranial fragment (MLD 1) from Makapansgat
Courtesy of Professor P. V. Tobias, photographed by A. R. Hughes

breccia. Subsequent search was rewarded by the recovery of the other half. When the specimens were developed from the matrix and restored they were found to constitute the major part of the brain-case, the base of the skull and part of the palate. The remainder of the facial skeleton and the frontal region is missing. The cranium has a remarkable resemblance to the best Sterkfontein skull (Sts 5) (*q.v.*), only differing in minor respects such as lack of an occipital torus, reduction of the post-glenoid process and in details of the cranial dimensions. The absence of an occipital torus has led to the suggestion that this specimen is probably female. The cranio-facial fragment discloses that the degree of prognathism was not large.

MANDIBLES
A range of specimens is available from this site displaying most of the mandibular morphology from infancy to senility. One of the best specimens is half an adult mandible (MLD 40) which contains the canine, premolar and molar teeth. The

Fig. 90   A juvenile mandible (MLD 2) from Makapansgat
Courtesy of Professor P. V. Tobias, photographed by A. R. Hughes

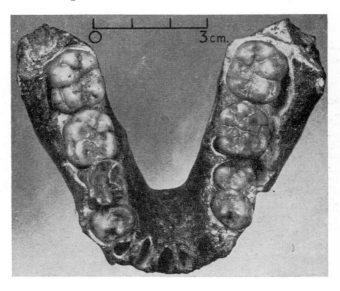

The Makapansgat Remains

body of the mandible is stout and on these grounds it is believed to be male. The symphysial region is damaged but the mental foramen is single. The ramus is short by comparison with the large australopithecine from Swartkrans, but comparable with those known from Sterkfontein.

TEETH

The principal morphological features of the teeth fall within the range of variation of those previously described from Sterkfontein.

POST-CRANIAL BONES

The Ischium (MLD 8) resembles the one that forms part of the Sterkfontein innominate, but differs markedly from the ischia of recent apes and from that of the Swartkrans australopithecine. It is a small bone with a moderate ischial tuberosity separated from the remains of the acetabular margin by a groove. This groove is narrower than that shown on the Sterkfontein pelvis and in this respect more nearly resembles the condition found in modern man.

The Ilia are adolescent, small and believed to be of opposite sex, the first specimen male (MLD 7) and the second female

Fig. 91   Two juvenile ilia (MLD 7 and MLD 25) and an ischial fragment (MLD 8) from Makapansgat
Courtesy of Professor P. V. Tobias, photographed by A. R. Hughes

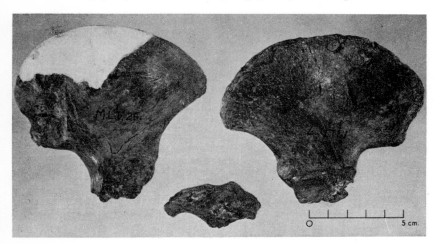

264

The Makapansgat Remains

(MLD 25). The blade of the ilium is broad and twisted into an S-shape when viewed from above. The anterior superior spine is prominent and reaches well forward, recalling the shape of the Swartkrans pelvis. The iliac pillar, which runs upwards from the acetabulum to the tuberosity of the iliac crest, is absent, but some thickening of the cortical bone in this region was disclosed when one specimen was accidently broken.

Although these pelvic fragments differ in a number of respects from their modern human counterparts they differ even more widely from the pelvic bones of modern apes. It is difficult to escape the conclusion that these australopithecines were habitually erect and bipedal in their form of locomotion.

The shaft of the *Humerus* is robust and well marked by the attachment of muscles. It is not unlike that of modern man (Boné, 1955). The remaining post-cranial bones are fragmentary.

Dimensions SKULL
Dart (1962a) *Holloway* (1970)
Max. Cranial Breadth 106 Cranial Capacity 435 cc
MANDIBLE (MLD 40)
Dart (1962b)
Max. Length 123·5 (damaged)
Body Height (M1) 36·0
Body Thickness (M1) 23·5
Max. Ramus Width 45·5
Max. Ramus Height (coronoid) 62·5
TEETH
Robinson (1956), *Dart* (1962b)

Upper Permanent Teeth (Crown Dimensions)								
	I1	I2	C	PM1	PM2	M1	M2	M3
Average l	—	—	—	8·8	9·4	12·5	13·9	—
figures b	—	—	—	12·0	12·6	12·8	15·7	—

MLD 40	Lower Permanent Teeth (Crown Dimensions)							
Left l	—	—	8·3*	10·0	10·0	12·8	15·0	15·0
side b	—	—	9·5*	11·0	14·0	12·3	14·3	14·0

* Damaged

The Makapansgat Remains

Affinities The affinities of the Makapansgat remains will be discussed with the rest of the australopithecine material.

Originals The Department of Anatomy, Medical School, University of the Witwatersrand, Johannesburg, Republic of South Africa.

Casts The University Museum, University of Pennsylvania, Philadelphia 4, Pennsylvania, U.S.A. (Cranio-facial fragment, adolescent mandible, ischium, ilia, calvarial and maxillary fragments.)

References DART, R. A. 1948a
The Makapansgat proto-human *Australopithecus prometheus*. *Am. J. Phys. Anthrop.* 6, 259-284.
DART, R. A. 1948b
The adolescent mandible of *Australopithecus prometheus*. *Ibid.*, 6, 391-412.
DART, R. A. 1949a
The cranio-facial fragment of *Australopithecus prometheus*. *Ibid.*, 7, 187-214.
DART, R. A. 1949b
Innominate fragments of *Australopithecus prometheus*. *Ibid.*, 7, 301-333.
DART, R. A. 1949c
A second adult palate of *Australopithecus prometheus*. *Ibid.*, 7, 335-338.
MAYR, E. 1950
Taxonomic categories in fossil hominids. *Cold Spring Harbour Symposia on Quantitative Biology* 15, 109-118.
KING, L. C. 1951
The geology of Makapan and other caves. *Trans. Roy. Soc. S. Afr.* 33, 121-151.
DART, R. A. 1954
The second or adult female mandible of *Australopithecus prometheus*. *Am. J. Phys. Anthrop.* 12, 313-343.
OAKLEY, K. P. 1954
Dating the australopithecines of Africa. *Ibid.*, 12, 9-23.
ROBINSON, J. T. 1954
The genera and species of the *Australopithecinae*. *Ibid.*, 12, 181-200.
BONÉ, E. L., and DART, R. A. 1955
A catalogue of australopithecine fossils found at the Limeworks, Makapansgat. *Ibid.*, 13, 621-624.
DART, R. A. 1955
Australopithecus prometheus and *Telanthropus capensis*. *Ibid.*, 13, 67-96.

HOWELL, F. C. 1955
The age of the australopithecines of Southern Africa. *Ibid.*, *13*, 635-662.

ROBINSON, J. T. 1956
The dentition of the *Australopithecinae*. *Transv. Mus. Mem.* *9*, 1-179.

BRAIN, C. K. 1957
New evidence for the correlation of the Transvaal ape-man bearing cave deposits. *Proc. Third Pan-African Cong. Prehist.*, *Livingstone*, *1955*, 143-148. Ed. J. D. Clark. London: Chatto and Windus.

DART, R. A. 1957a
The osteodontokeratic culture of *Australopithecus prometheus*. *Transv. Mus. Mem.* *10*, 1-105.

DART, R. A. 1957b
The second adolescent (female) ilium of *Australopithecus prometheus*. *J. palaeont. Soc. India 2*, 73-82.

EWER, R. F. 1957
Faunal evidence on the dating of the *Australopithecinae*. *Proc. Third Pan-African Cong. Prehist.*, *Livingstone*, *1955*, 135-142. Ed. J. D. Clark. London: Chatto and Windus.

OAKLEY, K. P. 1957
Dating the australopithecines. *Ibid.*, 155-157.

BRAIN, C. K. 1958
The Transvaal ape-man bearing cave deposits. *Transv. Mus. Mem.* *11*, 1-125.

DART, R. A. 1958
A further adolescent australopithecine ilium from Makapansgat. *Am. J. Phys. Anthrop. 16*, 473-480.

DART, R. A. 1959
The first australopithecine cranium from the pink breccia at Makapansgat. *Ibid.*, *17*, 77-82.

DART, R. A. 1962a
The Makapansgat pink breccia australopithecine skull. *Ibid.*, *20*, 119-126.

DART, R. A. 1962b
A cleft adult mandible and nine other lower jaw fragments from Makapansgat. *Ibid.*, *20*, 267-286.

KURTÉN, B. 1962
The relative ages of the australopithecines of Transvaal and the pithecanthropines of Java. In *Evolution und Hominisation*, 74-80. Ed. G. Kurth. Stuttgart: Gustav Fischer Verlag.

COOKE, H. B. S. 1964
Pleistocene mammal faunas of Africa, with particular reference to South Africa. In *African ecology and human evolution*, 65-116. Eds. F. C. Howell and F. Bourlière. London: Methuen and Co. Ltd.

COOKE, H. B. S. 1970
Notes from Members: Dalhousie University, Halifax, Canada. *News Bull. Soc. Vertebr. Paleont. 90*, 2.

The Makapansgat Remains

HOLLOWAY, R. L. 1970
New endocranial values for the australopithecines. *Nature* 227, 199–200.

WOLBERG, D. L. 1970
The hypothesized osteodontokeratic culture of the Australopithecinae: a look at the evidence and the opinions. *Curr. Anthrop.* 11, 23–30.

DART, R. A. 1971
On the osteodontokeratic culture of the Australopithecinae. *Curr. Anthrop.* 12, 233–235.

WOLBERG, D. L. 1971
Curr. Anthrop. Comment. Reply. *Curr. Anthrop.* 12, 235–236.

PARTRIDGE, T. C. 1973
Geomorphological dating of cave openings at Makapansgat, Sterkfontein, Swartkrans and Taung. *Nature* 246, 75–79.

BUTZER, K. W. 1974
Curr. Anthrop. Comment. Reply. *Curr. Anthrop.* 15, 413–416.

The announcement of the discovery of the Taung skull by Professor R. A. Dart in 1925, and the creation of a new genus in which to place it, provoked widespread controversy because he firmly emphasized the hominid features of the skull and dentition. Those opposed to Dart's assessment regarded these hominid features as being due to parallel evolution and of little significance. The principal grounds for doubt were reinforced by the fact that the specimen was juvenile. Such a small cranial capacity and such ape-like facial features, in an infant, would indicate a small adult brain size and grossly anthropoid facial features on pedomorphic grounds. None the less, the presence of a first permanent molar of undoubted hominid form should, in retrospect, have cautioned Dart's opponents.

Between 1936 and 1949, Dr Robert Broom, who had supported Dart's opinion, discovered three sites in the Transvaal —Sterkfontein, Kromdraai and Swartkrans—from which, with the assistance of Dr J. T. Robinson, he recovered a great deal of material including skulls, several hundreds of teeth and a number of post-cranial bones. Following Broom's death in 1951, more material was recovered by Robinson. Meanwhile farther north at Makapansgat, Dart was obtaining many new specimens of the same type from similar deposits.

Following the study of these remains it has become clear that the material can be divided into two principal groups. These groups comprise the remains of a small, generalized, light-framed form represented from Taung, Sterkfontein and Makapansgat, and the remains of a larger, more dentally specialized, robust creature from Kromdraai and Swartkrans. The smaller form seems to be derived from deposits which are earlier than those containing the larger form.

The species represented by the Taung infant skull was named *Australopithecus africanus;* later discoveries of the smaller form were named *Plesianthropus transvaalensis* (Sterkfontein) and *Australopithecus prometheus* (Makapansgat).*

---

* The specific name *prometheus* was given because it was mistakenly believed that there was evidence of the use of fire in association with the fossils at Makapansgat.

It is now widely accepted that the remains from Taung, Sterkfontein and Makapansgat belong to one species only, *A. africanus*. There is less agreement over the classification and nomenclature of the larger form. Broom believed that the

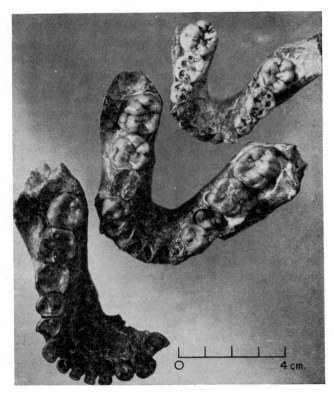

Fig. 92   A series of Australopithecine mandibles
Top—Taung
Middle—Makapansgat juvenile
Bottom—Makapansgat adult
Courtesy of Professor P. V. Tobias, photographed by A. R. Hughes

Kromdraai remains represented a distinct australopithecine genus and named the find *Paranthropus robustus;* later finds from Swartkrans were accepted as congeneric but of a new species *P. crassidens*. Robinson (1954 *et seq.*) has adhered to the generic separation of the larger and smaller forms, accepting *Paranthropus* as a genus containing one species and two sub-species in the Transvaal (*P. robustus robustus* and *P. robustus crassidens*), one species in East Africa (*P. boisei*) and another in the Far East (*P. palaeojavanicus*).

Other authorities regard all of the Transvaal material as representative of one genus containing two species, *Australopithecus africanus* and *Australopithecus robustus* (Oakley, 1954; Campbell, 1964), further distinctions being made between subspecies. Moreover, Mayr (1950) suggested that all the Transvaal material should be included within the genus *Homo*.

A further modification of this viewpoint has been put forward by Robinson (1972); he suggests that all of the more gracile australopithecine material should be placed in the genus *Homo* while retaining the genus *Paranthropus* for the robust remains.

Finally the 'single species hypothesis' of australopithecine taxonomy has been vigorously pursued by Wolpoff (1968, 1970, 1971; Wolpoff and Lovejoy, 1975). This viewpoint regards all of the australopithecine material as belonging to one species (*A. africanus*), the differences between robust and gracile forms being explained on the grounds of sexual dimorphism. This view has received little support.

The East African hominid remains from Olduvai (*q.v.*) and Koobi Fora (*q.v.*) have clarified matters a little but also raised other problems. It seems to be clear that australopithecines of robust type are now known from Olduvai as well as Koobi Fora and other sites in East Africa *i.e.* Olduvai Hominid 5, Koobi Fora KNM-ER 406, Peninj Kenya and Hadar, Ethiopia. It may be that gracile examples are also known from Olduvai (O.H. 24) and Koobi Fora (KNM-ER 1813) as well as Hadar (Al-288), living contemporaneously with recognizable early members of the genus *Homo* from each of these sites. If that proves to be the case, we must look even further back into the Plio-Pleistocene period in order to find the point of divergence of these hominid lines from their common ancestry. Meanwhile the Transvaal australopithecines provide morphological evidence of a 'prehuman' phase of hominid evolution in which considerable advance had been

The Affinities of the Australopithecines

made in the modification of the teeth for an omnivorous diet, the post-cranial skeleton for the development of upright posture and bipedal gait, whilst expansion of the brain had progressed but little.

Additional references GREGORY, W. K., and HELLMAN, M. 1939
The dentition of the extinct South African man-ape. *Australopithecus (Plesianthropus) transvaalensis* (Broom). *Ann. Transv. Mus. 19*, 339-373.
CLARK, W. E. LE GROS 1952
Hominid characters of the australopithecine dentition. *J. R. anthrop. Inst. 80*, 37-54.
TOBIAS, P. V. 1963
Cranial capacity of *Zinjanthropus* and other australopithecines. *Nature 197*, 743-746.
CLARK, W. E. LE GROS 1967
Man-ape or ape-men. New York: Holt, Rinehart and Winston, Inc.
WOLPOFF, M. H. 1968
'*Telanthropus*' and the single species hypothesis. *Am. Anthrop. 70*, 477-493.
WOLPOFF, M. H. 1970
The evidence for multiple hominid taxa at Swartkrans. *Am. Anthrop. 72*, 576-607.
WOLPOFF, M. H. 1971
Competitive exclusion among Lower Pleistocene hominids: the single species hypothesis. *Man 6*, 601-614.
ROBINSON, J. T. 1972
Early hominid posture and locomotion. Chicago: Chicago University Press.
WOLPOFF, M. H. and LOVEJOY, C. O. 1975
A rediagnosis of the genus *Australopithecus*. *J. Hum. Evol. 4*, 275-276.

The Saldanha Skull

Fig. 93   The Saldanha calvaria
Right lateral view
Courtesy of Professor R. Singer

Synonyms and other names	*Homo saldanensis* (Drennan, 1955); *Homo sapiens rhodesiensis* (Campbell, 1964) Saldanha man; Hopefield man
Found by	K. Jolly and R. Singer, 1953.
Site	Elandsfontein Farm, 10 miles south-west of Hopefield and 15 miles south-east of Saldanha Bay, Cape Province, Republic of South Africa.
Geology	The remains were found on the surface of sandy veld, 300 feet above sea level. The fossil horizon is made from a nodular calcrete representing a dried-out pan-floor in which the bones had accumulated. Ridges of ferricrete cut across the site and indicated previous wetter conditions, but the fossil layer is capped by surface limestones produced during drought (Mabbutt, 1956 and 1957). A reappraisal of the geology of the site has been given recently (Butzer, 1973). Much of the earlier work has been confirmed but much detail has been added.

273

The Saldanha Skull

Associated finds The stone tools which were found in great numbers are of three principal types (Singer, 1954; Singer and Crawford, 1958).

1. *Early Stone Age* (final Acheulian or Modified Fauresmith), tools including cleavers, large and pygmy hand-axes, bolas-like stones, pebble choppers and unconventional tools. They were made of silcrete, quartzite, felspar or soft sandstone. The skull bones were associated with this industry.

2. *Middle Stone Age* tools (Stillbay), made of fine-grained ferruginous silcrete.

3. *Later Stone Age* tools made of felspar, sandstone and silcrete. All of these rocks, with the exception of felspar, are foreign to the site.

Supposed bone 'chisels' and 'gouges' reported from the site have been discounted as carnivore chewed and weathered bones (Singer, 1956). Recent appraisal of the radiocarbon chronology of the South African stone age suggests that the Middle Stone Age may have ended at 30–40,000 years B.P. and that the Early Stone Age/Middle Stone Age transition junction may predate radiocarbon limits (Vogel and Beaumont, 1972; Klein, 1973). Thus the junction may be in the Middle Pleistocene.

The fossil fauna recovered from the site includes elephant (*cf. Loxodonta atlantica*), a baboon (*Papio ursus*), a large primate (*cf. Simopithecus sp.*), rhinoceros (*cf. Diceros bicornis* and *cf. Ceratotherium simum*), horse (*cf. Equus plicatus*), giant pig (*Mesochoerus lategani*), hippopotamus (*Hippopotamus amphibius*), long-horned buffalo (*Homioceras baini*) and several other artiodactyls and carnivores (Cooke, 1964).

A faunal list has been given by Boné and Singer (1965), but since then a number of additions to and revisions of the list have been published (Hendey, 1969; Cooke, 1973; Klein, 1973). The consensus of these views is that there are similarities between the Elandsfontein assemblage, the Vaal-Cornelia assemblage and that from Olduvai, Bed IV. It may be therefore, that the Elandsfontein fauna is late Middle Pleistocene rather than early Upper Pleistocene as formerly believed.

Dating Initially this site was considered on geological, archaeological and faunal evidence to be of early Upper Pleistocene age. 274

The Saldanha Skull

Recent reconsiderations of the fauna and the artefacts (*see above*) suggest that this site should be referred to the late Middle Pleistocene. Radiometric methods have not proved of much value at this site up to the present.

Morphology Initially 11 fragments were recovered loose on the site, and subsequently more pieces including a portion of mandible. The fragments were fitted together and the cranial vault was finally reconstructed from 27 pieces of bone.

CALVARIA
The cranial vault is low having a flattened retreating forehead and massive supra-orbital ridges. The parietals are gently rounded but the greatest breadth of the skull must have been near its base. The occipital torus is prominent and the supreme nuchal line is well inscribed; there is no occipital bun-formation. The position of the foramen magnum is not known and the nuchal plane is only partly represented (Singer, 1954). Drennan (1953a and b, 1955) believed that the occipital bone, and thus the nuchal plane, to have been tilted backwards as in the Solo skulls. On these grounds he suggested a crouching posture and a more primitive status for this form, a suggestion dismissed by Singer (1957) as pure conjecture in the absence of most of the occipital bone, the auditory meatus, the orbit and therefore of the Frankfurt Plane orientation.

THE MANDIBLE
The fragment of mandible consists of a portion of the right ramus anterior to the inferior dental canal. Its size and shape indicates that the ramus was broad and the mandibular notch shallow. The fragment is remarkable in its resemblance to the corresponding portion of the Heidelberg jaw.

Dimensions CALVARIA
Maximum Length 200   Maximum Breadth 144
Cranial Index 72·0 (Dolichocephalic)
Cranial Capacity 1,200–1,250 cc (Drennan, 1953a)
            Calculated 1,250 cc (Drennan, 1953b)
MANDIBULAR FRAGMENT
Alveolar Plane-Coronoid Height 40 (as in the Heidelberg jaw)

The Saldanha Skull

Affinities　From the beginning the similarity of the Saldanha skull and the Rhodesian skull was recognized, but the morphological identity of the two specimens was not accepted (Drennan, 1953a and b). Where differences occurred the Saldanha skull was said to resemble the Solo skulls, thus Saldanha man was believed to be a more primitive form of Rhodesian man. It was concluded by Singer (1954) that Saldanha and Rhodesian man were African Neanderthalians, unlike the European but similar to the Asiatic representatives of this group, i.e. Solo man. Later Drennan (1955) created a new specific name for Saldanha man, *Homo saldanensis*; this step was rejected by Singer (1958) who stated that the Saldanha and Rhodesian skulls are similar in the parts that are there to compare, and that their differences are within sex or normal variability limits.

However, it is gradually becoming accepted that both Rhodesian and Saldanha man are of the same subspecies of *Homo sapiens*, *Homo sapiens rhodesiensis* (Campbell, 1964).

Originals　South African Museum, Cape Town, Republic of South Africa.

Casts　The University Museum, University of Pennsylvania, Philadelphia 4, Pennsylvania, U.S.A. (Calvarium only).

References　DRENNAN, M. R. 1953a
A preliminary note on the Saldanha skull. *S. Afr. J. Sci. 50*, 7-11.
DRENNAN, M. R. 1953b
The Saldanha skull and its associations. *Nature 172*, 791-793.
SINGER, R. 1954
The Saldanha skull from Hopefield, South Africa. *Am. J. Phys. Anthrop. 12*, 345-362.
DRENNAN, M. R. 1955
The special features and status of the Saldanha skull. *Am. J. Phys. Anthrop. 13*, 625-634.
DRENNAN, M. R., and SINGER, R. 1955
A mandibular fragment, probably of the Saldanha skull. *Nature 175*, 364-365.
MABBUTT, J. A. 1956
The physiography and surface geology of the Hopefield fossil site. *Trans. roy. Soc. S. Afr. 35*, 21-58.
SINGER, R. 1956
The 'bone tools' from Hopefield. *Amer. Anthrop. 58*, 1127-1134.
276

The Saldanha Skull

MABBUTT, J. A. 1957
The physical background to the Hopefield discoveries. *Proc. Third Pan-African Cong. Prehist., Livingstone, 1955,* 68-75. Ed. J. D. Clark. London: Chatto and Windus.

OAKLEY, K. P. 1957
The dating of the Broken Hill, Florisbad and Saldanha skulls. *Ibid.* 76-79.

SINGER, R. 1957
Investigations at the Hopefield site. *Ibid.,* 175-182.

SINGER, R. 1958
The Rhodesian, Florisbad and Saldanha skulls. In *Hundert Jahre Neanderthaler,* 52-62. Ed. G. H. R. von Koenigswald. Utrecht: Kemink en Zoon.

SINGER, R., and CRAWFORD, J. R. 1958
The significance of the archaeological discoveries at Hopefield, South Africa. *J. R. Anthrop. Inst. 88,* 11-19.

CAMPBELL, B. 1964
Quantitative taxonomy and human evolution. In *Classification and human evolution,* 50-74. Ed. S. L. Washburn. London: Methuen and Co. Ltd.

BONÉ, E. and SINGER, R. 1965
Hipparion from Langebaanweg, Cape Province, and a revision of the genus in Africa. *Ann. S. Afr. Mus. 48,* 273-397.

SINGER, R. and WYMER, J. 1968
Archaeological investigations at the Saldanha Skull Site in South Africa. *S. Afr. Archaeol. Bull. 23,* 63-74.

HENDEY, Q. B. 1969
Quaternary vertebrate fossil sites in the southern Cape Province. *S. Afr. Archaeol. Bull. 24,* 96-105.

VOGEL, J. C. and BEAUMONT, P. B. 1972
Revised Radiocarbon Chronology for the Stone Age in South Africa. *Nature 237,* 50.

BUTZER, K. W. 1973
Re-evaluation of the geology of the Elandsfontein (Hopefield) site, South-western Cape, South Africa. *S. Afr. J. Sci. 69,* 234-238.

KLEIN, R. C. 1973
Geological antiquity of Rhodesian Man. *Nature 244,* 311-312.

COOKE, H. B. S. 1973
The fossil mammals of Cornelia. O.F.S. South Africa. *Mem. Nat. Mus. (S. Afr.).*

The Florisbad Skull

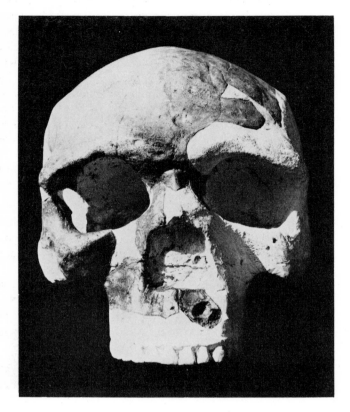

Fig. 94   A reconstruction of the Florisbad skull.
Reconstructed by T. F. Dreyer
Courtesy of the Director of the Nasionale Museum,
Bloemfontein

Synonyms and other names	Homo (*Africanthropus*) *helmei* (Dreyer, 1935); *Homo florisbadensis* (*helmei*) (Drennan, 1935); *Homo sapiens* (Vallois, 1957)
Found by	T. F. Dreyer, 1932.
Site	Florisbad, 30 miles north of Bloemfontein, Orange Free State, Republic of South Africa.

278

The Florisbad Skull

Geology   The deposits are on the southern slope of the Hagenstad salt-pan which is near a medicinal watering place. The area is marked by numerous springs, many of which have become choked with accumulated debris; when this occurs a new spring eye opens near by. The debris consists of sand containing stone artefacts, broken bones and teeth. The heavier sand is ilmenite with garnets and diopside, whilst above this layer there is a cap of pure white quartz sand (Dreyer, 1935).

Recent investigation of the site has disclosed a profile of eleven strata including a basal layer and four other layers of 'peat'. These 'peat' layers are in fact dark coloured sand and clay containing little organic matter.

The skull was found at the side of the eye of a spring beneath the 'Peat I' stratum about 18 feet from the surface.

Associated finds   Dreyer reported the occurrence of stone tools of African Middle Stone Age culture. In particular one group of Mousterian-like blades known as the Hagenstad variation was recovered from the deeper layers.

The fossil mammalian fauna that was found includes several living rodents and carnivores as well as extinct species such as giant buffalo (*Connochaetes antiquus, Alcelaphus helmei*) and two equids (*Equus helmei, Equus burchelli*) (Cooke, 1964).

Dating   The stratigraphy of this specimen is confusing from the point of view of establishing its age. However, the implements suggest that the deposits belong to the Upper Pleistocene or even the Holocene. A pollen investigation (van Zinderen Bakker, 1957) suggests that the oldest parts of the profile were probably formed at the beginning of the Upper Pleistocene during a dry phase or interpluvial. The radiocarbon dates of 41,000 years B.P. given for the 'Dark Layer I' by Libby (1954) and of 37,000 years B.P. (quoted by van Zinderen Bakker, 1957) have been considered too high because the Pleistocene plant material in that layer is probably contaminated by 'dead carbon' carried up by the spring from underlying Palaeozoic coal measures (Oakley, 1957).

New relative and absolute dating of this skull has been given recently (Protsch and Goethe, 1974).   279

The Florisbad Skull

Morphology   The skull fragment consists of part of the face and vault including the right orbital margin and part of the maxilla; the base of the skull and the mandible are missing. The cranium is large but rather flattened with no parietal, and feeble frontal, bosses. The superciliary ridges are as those found in modern man, there being no evidence of a supra-orbital torus. The face is moderately prognathic. The palate is incomplete and the only tooth present is the right upper third molar which is very worn.

Dimensions   Max. Length approx. 200   Max. Breadth approx. 150.

Affinities   In the original description of this skull Dreyer (1935) named the specimen *Homo (Africanthropus) helmei* thus asserting that this form was sub-generically distinct from other members of the genus *Homo*. However, in the same publication Kappers, who had studied the endocranial cast, emphasized its likeness to *Homo sapiens fossilis*. Drennan (1935, 1937) urged the Neanderthal characters of this skull and endocast on metrical grounds, and proposed the name *Homo florisbadensis (helmei)* as being more appropriate. This view was opposed by Galloway (1937, 1938) who emphasized that the non-metrical features of the Florisbad skull linked it with *Homo sapiens*, and moreover with the Australoid variety of this species. Similarly Boule and Vallois (1957) unequivocally classified these remains as belonging to *Homo sapiens*.

Original   National Museum, Bloemfontein, Orange Free State, Republic of South Africa.

Casts   The University Museum, University of Pennsylvania, Philadelphia 4, Pennsylvania, U.S.A.

References   DREYER, T. F. 1935
A human skull from Florisbad. *Proc. Acad. Sci. Amst. 38*, 119-128.
DRENNAN, M. R. 1935
The Florisbad skull. *S. Afr. J. Sci. 32.* 601-602.
DREYER, T. F. 1936
The endocranial cast of the Florisbad skull—a correction. *Sool. Nav. nas. Mus. Bloemfontein 1*, 21-23.
DRENNAN, M. R. 1937
The Florisbad skull and brain cast. *Trans. roy. Soc. S. Afr. 25*, 103-114.
GALLOWAY, A. 1937
Man in Africa in the light of recent discoveries. *S. Afr. J. Sci. 34*, 89-120.

The Florisbad Skull

GALLOWAY, A. 1938
The nature and status of the Florisbad skull as revealed by its non-metrical features. *Am. J. Phys. Anthrop. 23*, 1-16.

LIBBY, W. F. 1954
Chicago radiocarbon dates, V. *Science 120*, 733-742.

BAKKER, E. M. VAN ZINDEREN 1957
A pollen analytical investigation of the Florisbad deposits (South Africa). *Proc. Third Pan-African Cong. Prehist., Livingstone, 1955,* 56-57. Ed. J. D. Clark. London: Chatto and Windus.

OAKLEY, K. P. 1957
The dating of the Broken Hill, Florisbad and Saldanha skulls. *Ibid.,* 76-79.

BOULE, M., and VALLOIS, H. V. 1957
Fossil man, 4th Ed., 462. London: Thames and Hudson.

SINGER, R. 1958
The Rhodesian, Florisbad and Saldanha skulls. In *Hundert Jahre Neanderthaler,* 52-62. Ed. G. H. R. von Koenigswald. Utrecht: Kemink en Zoon.

COOKE, H. B. S. 1964
Pleistocene mammal faunas of Africa, with particular reference to Southern Africa. In *African ecology and human evolution,* 65-116. Eds. F. C. Howell and F. Bourlière. London: Methuen and Co. Ltd.

PROTCH, R. 1973
The dating of Upper Pleistocene subsaharan fossil hominids and their place in human evolution: the morphological and archaeological implications. *Ph.D. thesis, University of California, L.A.*

The Far East

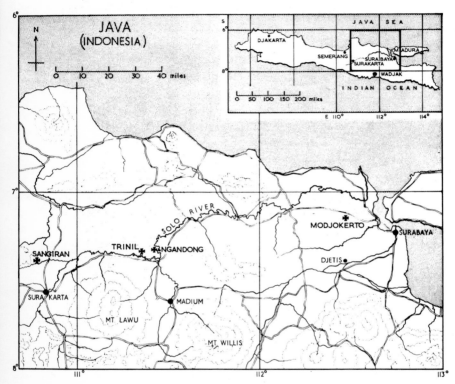

Fig. 95   Hominid fossil sites in Java

The Trinil Remains

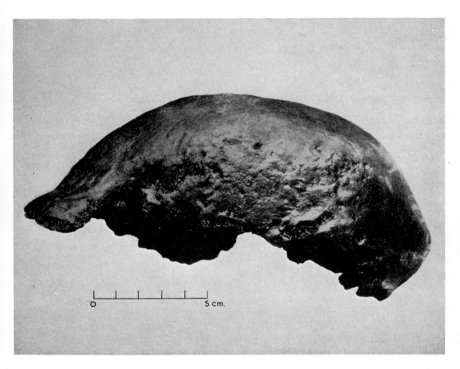

Fig. 96   The Trinil calotte (Pithecanthropus I)
Left lateral view
Courtesy of Professor J. S. Weiner and Dr D. A. Hooijer

Synonyms
and other names   Anthropopithecus erectus (Dubois, 1892); Pithecanthropus erectus (Dubois, 1894); Pithecanthropus I (von Koenigswald and Weidenreich, 1939); Homo erectus javensis (Weidenreich, 1940); Homo erectus (Mayr, 1950); Homo erectus erectus (Dobzhansky, 1944; Campbell, 1964); Trinil 1–9 (Oakley, Campbell and Molleson, 1975); Java man.

Site   Trinil, approximately 20 miles north-west of Madium, six miles west of Ngawi, Central Java, Indonesia.

Found by   Eugene Dubois, 1891–1898; G. Kriele, 1900.

Geology   Trinil lies at the foot of a volcano, Lawu, whose lavas and cinders have spilled over a wide area. Elevated Pleistocene   285 deposits include the Kabuh beds which overlie the Putjangan beds. The Trinil hominid remains were found in the Kabuh beds which consist of fresh-water sandstones and conglomerates containing volcanic material, within a few yards of the waters of the river Solo. The femur was uncovered 15 metres upstream.

Associated finds No artefacts were found in association with the remains, but numerous fossil mammalian bones have been recovered from the site. Amongst these were the remains of stegodont elephant (*Stegodon*), rhinoceros (*Rhinoceros sondaicus*), carnivores (*Felis*), and some ungulates including axis deer (*Axis leydekkeri*) and a small antelope (*Duboisia kroesenii*) (Selenka and Blanckenhorn, 1911). This fauna is now known as the Trinil fauna and is typical of the Kabuh beds.

Dating According to von Koenigswald (1934, 1949), three series of deposits succeeded each other in Central Java; the Djetis (Lower Pleistocene), the Trinil (Mid-Pleistocene) and the Ngandong (Upper Pleistocene) with a more recent fauna. Hooijer (1951, 1956, 1957, 1962) has accepted the Trinil bed as Mid-Pleistocene, but has denied a faunal distinction between the Djetis and Trinil layers and claims that they are probably both Middle Pleistocene. Estimations of the fluorine content of the Trinil calotte and femur have established their contemporaneity both with each other and with their associated fauna (Bergman and Karsten, 1952). In an attempt to correlate the dating of European, African and Asian fossils Kurtén (1962) has equated *Pithecanthropus erectus* of Java with *Paranthropus* from South Africa, and assigned to them both a Middle Pleistocene date (Mindel or Antepenultimate Glaciation). Later by the use of the radiometric potassium–argon dating method of tektites and basalt found in deposits which are said to correspond with those at Trinil, von Koenigswald et al. (1962) have claimed a chronological age of 550,000 years B.P. for *Pithecanthropus erectus*, within the First Glaciation (Günz or Early Glaciation). Subsequently a potassium/argon date of approximately 500,000 years B.P. was obtained from Trinil deposits related to the Muriah volcano and a date of over 700,000 years B.P. from tektites recovered from Kabuh Beds (Trinil fauna) at Sangiran (Von Koenigswald, 1968). Recent analytical studies

The Trinil Remains

on Trinil fossils neither confirms nor denies the Middle Pleistocene antiquity of the material although it probably confirms the provenance of the calotte and femora I–V (Trinil 3, 6, 7, 8 and 9) (Day and Molleson, 1973).

Morphology The finds include a calotte made up of parts of the frontal, parietal and occipital bones with little or no sign of suture lines. It is thick, undistorted and heavily mineralized. The frontal region of the vault is markedly flattened in profile, leading forwards to a heavy supra-orbital torus which is hollowed by paired frontal air sinuses. Behind the brow ridge the frontal region is sharply constricted producing a post-orbital waist, whilst in the midline the bone is heaped into a sessile ridge or keel. The temporal lines are well shown but widely separated. Internally the calotte is moulded by the cerebral convolutions and grooved by the meningeal blood vessels.

In addition to the calotte some femoral remains were recovered. They include one complete femur and four

Fig. 97 Trinil femora.
Left: right femur (2) Posterior view
Right: left femur (1) Anterior view
Photographed by courtesy of the Director of the Rijksmuseum von Natuurlijke Histoire, Netherlands

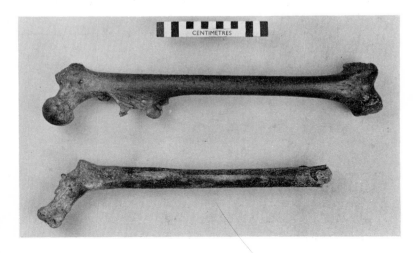

287

The Trinil Remains

femoral fragments. A sixth femoral fragment recovered from Kedung Brubus has recently been discounted (Day and Molleson, 1973).

The complete femur is remarkable in its general resemblance to that of modern man. The shaft is straight and has a prominent linea aspera, but in its upper third the specimen is marred by a pathological outgrowth. The head of the bone is rounded, the neck stout and angulated to the shaft, whilst the greater and lesser trochanters are well developed indicating the attachment of powerful muscles around the hip joint.

The features of this femur, in particular the 'weight-carrying-angle' between the shaft and the condyles, suggest strongly that Java man was capable of standing and walking erectly. A recent study of the Trinil femora (Day and Molleson, 1973) concludes that the gross, radiological and microscopical anatomy of these bones does not distinguish them from modern human femora.

Dimensions CALOTTE (TRINIL 2)
Length 183   Breadth 134
Cephalic Index 70 (Dolichocephalic)
Cranial capacity 900 cc (Dubois, 1924)
          914 cc (von Koenigswald and Weidenreich, 1939)
          850 cc (Boule and Vallois, 1957)
          940 cc (Ashley Montagu, 1960)

FEMUR I (TRINIL 3)
Length 455   Mean mid-shaft diameter 28·6
Bicondylar breadth 77   Mean diameter of head 44·8
Angle of neck 122°   Femoro-condylar angle 100°

Affinities These were the first significant finds in the search for fossil man in the Far East. Despite the opposition of Dubois, subsequent finds at Sangiran (*q.v.*) and Modjokerto confirmed that the Trinil specimens are representative of a group of hominids who occupied Java in the Middle Pleistocene. Further finds in China (Peking man) suggested that hominds of this grade were widely distributed in the Far East, for the morphological

288

The Trinil Remains

differences between the two groups indicate only racial variation (Weidenreich, 1938, 1940). Recently Java man has been classified as *Homo erectus erectus*, only subspecifically distinct from *Homo erectus pekinensis* (Dobzhansky, 1944; Campbell, 1964).

Originals Collection Dubois, Rijksmuseum von Natuurlijke Historie, Leiden, Netherlands.

Casts 1. Rijksmuseum von Natuurlijke Historie, Leiden, Netherlands.
2. The University Museum, University of Pennsylvania, Philadelphia 4, Pennsylvania U.S.A.

References DUBOIS, E. 1892
Palaeontologische anderzoekingen op Java. *Versl. Mijnw. Batavia 3,* 10–14.
DUBOIS, E. 1894
Pithecanthropus erectus, *eine menschenaehnliche Ubergangsform aus Java.* Batavia: Landesdruckerei.
DUBOIS, E. 1895
Pithecanthropus erectus du Pliocene de Java. P. V. Bull. Soc. belge Geol. 9, 151–160.
SELENKA, M. L., and BLANCKENHORN, M. 1911
Die Pithecanthropus—*Schichten auf Java.* Leipzig: Verlag von Wilhelm Engelmann.
DUBOIS, E. 1924
On the principal characters of the cranium and the brain, the mandible and the teeth of *Pithecanthropus erectus. Proc. Acad. Sci. Amst. 27,* 265–278.
DUBOIS, E. 1926
On the principal characters of the femur of *Pithecanthropus erectus. Proc. Acad. Sci. Amst. 29,* 730–743.
WEINERT, H. 1928
Pithecanthropus erectus. Z. ges. Anat. 87, 429–547.
KOENIGSWALD, G. H. R. VON 1934
Zur Stratigraphie des javanischen Pleistocän. *Ing. Ned. Ind. 1,* 185–201.
WEIDENREICH, F. 1938
Pithecanthropus and *Sinanthropus. Nature 141,* 378–379.
WEIDENREICH, F. 1940
Some problems dealing with ancient man. *Am. Anthrop. 42,* 375–383.
DOBZHANSKY, T. 1944
On the species and races of living and fossil men. *Am. J. Phys. Anthrop. 2,* 251–265.

The Trinil Remains

KOENIGSWALD, G. H. R. VON, 1949
The discovery of early man in Java and Southern China. In *Early man in the Far East*. Ed. W. W. Howells; Philadelphia. *Stud. Phys. Anthrop. 1*, 83-98.

MAYR, E. 1950
Taxonomic categories in fossil hominids. *Cold Spring Harbour Symposia on Quantitative Biology 15*, 109-118.

HOOIJER, D. A. 1951
The geological age of *Pithecanthropus, Meganthropus* and *Gigantopithecus. Am. J. Phys. Anthrop. 9*, 265-281.

BERGMAN, R. A. M., and KARSTEN, P. 1952
The fluorine content of *Pithecanthropus* and other specimens from the Trinil fauna. *Proc. Acad. Sci. Amst. B. 55*, 1, 150-152.

HOOIJER, D. A. 1956
The lower boundary of the Pleistocene in Java and the age of *Pithecanthropus. Quarternaria 3*, 1, 5-10.

HOOIJER, D. A. 1957
The correlation of fossil mammalian faunas and the Plio-Pleistocene boundary in Java. *Proc. Acad. Sci. Amst. B. 60*, 1-10.

HOOIJER, D. A. 1962
The Middle Pleistocene fauna of Java. In *Evolution und Hominisation*, 108-111. Ed. G. Kurth. Stuttgart: Gustav Fischer Verlag.

KURTÉN, B. 1962
The australopithecines of Transvaal and the pithecanthropines of Java. *Ibid.*, 74-80.

KOENIGSWALD, G. H. R. VON 1962
Das absolute Alter des *Pithecanthropus erectus* Dubois. *Ibid.*, 112-119.

DAY, M. H. and MOLLESON, T. I. 1973
The Trinil femora. In *Human evolution*. Symp. S.S.H.B. Vol. 11. Ed. M. H. Day. London: Taylor and Francis.

OAKLEY, K. P., CAMPBELL, B. G. and MOLLESON, T. I. 1975
Catalogue of fossil hominids Part III: Americas, Asia, Australasia. London: Trustees of the British Museum (Natural History).

The Sangiran Remains

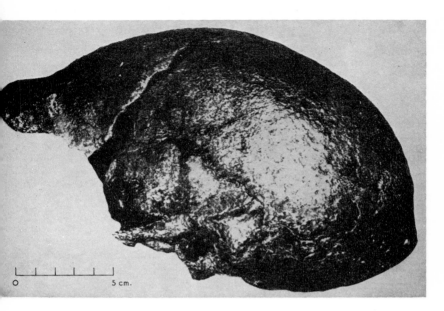

0   5 cm.

. 98 The first Sangiran calvaria (Pithecanthropus II)
Left lateral view
Courtesy of Professor J. S. Weiner

(In this section it is proposed that only the most significant
finds from this important site will be discussed. For a full
list of the remains, and their various designations, see table,
page 301).

Synonyms   1. *Pithecanthropus* (von Koenigswald, 1938); Pithecanthropus
1d other names   II (von Koenigswald and Weidenreich, 1939); *Homo erectus*
(Mayr, 1950); *Homo erectus erectus* (Campbell, 1964)
2. Pithecanthropus IV (von Koenigswald and Weidenreich
1939); *Pithecanthropus* (von Koenigswald, 1942); *Pithecan-*
thropus robustus (Weidenreich, 1945); *Pithecanthropus modjoker-*
tensis (von Koenigswald, 1950); *Homo erectus* (Mayr, 1950);
Pithecanthropus erectus (Piveteau, 1957); *Homo erectus erectus*
(Dobzhansky, 1944; Campbell, 1964)
3. Pithecanthropus VIII (Sartono, 1971)

291

4. *Meganthropus palaeojavanicus* (Weidenreich, 1945); Meganthropus I and II (Robinson, 1953); *Paranthropus palaeojavanicus*
(Robinson, 1954); *Pithecanthropus palaeojavanicus* (Piveteau,
1957); Meganthropus A (1941 and 1950 specimens) Meganthropus B (1952 specimen)

Site   Sangiran, by the river Tjemoro, a tributary of the Solo about
40 miles west of Trinil, near Surakarta, Central Java,
Indonesia.

Found by   1. G. H. R. von Koenigswald, September 1937.
2. Collectors employed by G. H. R. von Koenigswald,
1938–1939.
3. Mr Towikromo, a collector employed by S. Sartono,
13th September, 1969.
4. G. H. R. von Koenigswald and his assistant, 1941 and 1950
P. Marks, 1952.

Geology   Sangiran lies at the foot of a volcano, Lawu, to the north of
Surakarta. Lavas and ashes from this volcano have spilled
over a wide area creating deposits that are bedded. The
hominid finds from Sangiran have been recovered from the
upper Kabuh beds (Trinil fauna) and the lower Putjangan
beds (Djetis fauna). These beds are believed to be of Middle
Pleistocene and Early Pleistocene age respectively (Sartono,
1961). The Kabuh deposits consist of conglomerate, volcanic
tuffs and sandstones that show evidence of fluviatile deposition. The Putjangan bed consists of black fresh water clay-
stones with marine intercalations overlying lower volcanic
breccias. The provenance of the individual finds from
Sangiran is noted in Table 000. Further details of the geology
of the Sangiran area have been given by Sartono (1968, 1971
and 1972).

Associated finds   1. No artefacts were associated with the find but the fauna
from the Kabuh beds is known as the Trinil fauna and
includes the remains of stegodont elephant (*Stegodon*),
rhinoceros (*Rhinoceros sondaicus*), carnivores (*Felis*) and some
ungulates including axis deer (*Axis lydekkeri*) and a small
antelope (*Duboisia kroesenii*) (Selenka and Blankenhorn, 1911).
2. No artefacts were associated with this find but the fauna
from the Putjangan beds is known as the Djetis fauna and

includes a primitive horned ox (*Leptobos*) and a sabre-toothed
cat (*Epimachairodus*)
3. No artefacts were associated with this find which was
recovered from the Kabuh beds (Trinil fauna).
4. No artefacts were associated with these finds. The earlier
specimens were derived from the Putjangan beds (Djetis
fauna) and the most recent from the Kabuh beds (Trinil
fauna).

Dating The dating of the remains from Sangiran has been a difficult
problem for many years, being part of an even greater
problem, that of determining the precise age of all the Pleistocene deposits in Java. The sequence of deposits in east Java is
not generally disputed in that the Kabuh beds, which contain
the Trinil fauna are younger than the Putjangan beds
containing the Djetis fauna. The Sangiran area reveals both
sets of beds at several sites. In the past the Trinil and Djetis
faunas have been the subject of contention; Hooijer (1962)
has accepted the Trinil fauna as Middle Pleistocene and has
denied a distinction between this fauna and the Djetis fauna,
claiming that they are both probably Middle Pleistocene. On
the other hand von Koenigswald (1949) has always maintained
the view that the Djetis fauna is evidence of the Lower
Pleistocene age of the layer that contains these fossils.
Potassium-argon dating of the Kabuh beds has given dates
of 500,000 years B.P. and 830,000 years B.P. at differing sites
(von Koenigswald, 1964) and at Sangiran a tektite potassium-
argon estimation has given 710,000 years B.P. for the same
beds. A sample of pumice taken from the Putjangan formation at Modjokerto has been dated at 1·9 million years B.P. by
the potassium-argon method (Jacob, 1972) and suggests, by
correlation, that the Sangiran Putjangan beds may be of the
same age.

Morphology (1) The calvaria consists of the frontal, parietal, temporal and
occipital bones of an adult; both facial skeleton and skull base
are missing. The region of the foramen magnum and the
right side of the frontal bone are broken away. In general
form the specimen strongly resembles the Trinil calotte; the
flattening and keeling of the frontal region, the supra-orbital 293

torus and the widely separated temporal lines are features common to both. The occipital bone of the Sangiran calvaria shows that the nuchal plane is inclined at an angle intermediate between that of the pongids and modern man; the supramastoid and occipital crests are in continuity and the mastoid processes are very small. The foramen magnum is placed forward, evidence that *Pithecanthropus* was habitually upright in posture. Internally the bones are impressed by the cerebral convolutions and grooved by the larger dural venous sinuses and meningeal vessels.

Dimensions Length c. 180 Max. Breadth 140
Cephalic Index 77·8 (Mesocephalic)
Cranial Capacity c.750 cc (von Koenigswald, 1938)
          850 cc (Weidenreich, 1938)
          775 cc (von Koenigswald, 1949)
          815 cc (Boule and Vallois, 1957)

Comparisons of the dimensions of the Sangiran calvaria with those of the Trinil calotte led to the suggestion that the Sangiran specimen was female (von Koenigswald, 1938).

Morphology (2) The remains consists of the posterior half of a brain case and the lower portion of both maxillae.

CALVARIA
The cranium is represented by almost the entire occipital bone, including the foramen magnum and the occipital condyles, the temporal bones and approximately the posterior three-quarters of both parietal bones. The skull is larger than that of Pithecanthropus I or II, but resembles them in that it has a low vault and has its greatest breadth at the base. There is a marked frontal keel, accentuated by parasagittal depressions and a series of 'knob-like' processes leading back from the vertex to the occipital torus—features not seen in any other pithecanthropine. The occipital torus is very large and joins the supra-mastoid ridges on either side; the nuchal muscles have left well marked impressions on the occipital bone indicating a powerful neck. The mastoid processes are large and project downwards and inwards, in contrast to the small processes of the first Sangiran calvaria. The external

294

Fig. 99   The Sangiran maxilla (Pithecanthropus IV)
Right lateral view
Courtesy of Professor G. H. R. von Koenigswald

auditory meatus is oval as in modern man, but the tympanic
plate is thick; the mandibular fossa is deep and narrow and
the articular eminence is absent. Internally the petrous
temporal bone is very prominent.

MAXILLAE

The alveolar processes are complete except for the posterior
part of the left side. In addition almost the entire palate, the
floor of the nasal cavity and the maxillary sinuses of both
sides are preserved.

The maxillae were crushed before fossilization, resulting in
some distortion. The bony palate is very large and smooth,
relieved only by the presence of the palatal groove on the right
side which is limited medially by an unusual bony prominence. The incisive canal is double and has a funnel-like   295

opening which is very large and distally placed. The pre-maxillary region is very deep but the pillars of the zygomatic bones arise near the alveolar borders. In lateral view the degree of facial and alveolar prognathism is large but there is no typical nasal spine. The maxillary sinus is large but does not extend back into the maxillary tuberosity.

TEETH

All the teeth are present except the incisors and the left second and third molars. An isolated incisor was found with the previous specimens. The teeth are very little worn. The incisor sockets indicate that the teeth sloped forwards, and that a wide diastema existed between the lateral incisor and the canine. The canines are large by comparison with hominid teeth and their breadths exceed their lengths. The molars decrease in size in the order $M_2 > M_1 > M_3$ and the cusp pattern of their crowns does not differ appreciably from those of Peking man except that the remains of the cingulum are less obvious in the Sangiran molars.

Dimensions   *Weidenreich* (1954)

CALVARIA

Max. Length 199   Max. Breadth ?158
Cranial Index ?79·3   Cranial Capacity c. 900 cc

				Upper Teeth (Crown Dimensions)					
Pith. IV		I^1	I^2	C	PM^1	PM^2	M^1	M^2	M^3
Left	l	—	—	9·5	8·5	8·5	12·3	—	—
side	b	—	—	11·9	12·4	12·3	13·6	—	—
Right	l	—	10·0	9·5	8·2	8·2	12·1	13·6	10·8
side	b	—	10·4	11·7	12·4	12·1	13·7	15·2	14·0

Morphology (3)   Pithecanthropus VIII (Sangiran 17) is the most complete skull recovered from this site, indeed the best preserved specimen that has emerged from Java so far. The vault of the skull is almost complete and the face is intact apart from the loss of the left zygomatic region. The base is also well preserved posteriorly but the body of the sphenoid is broken away. The foramen magnum is present but a little broken at its rim. The palate is present in large part and a number of teeth are present including the three molars on the right as well as the canine, while on the left the second premolar is present. The face has suffered some distortion and is displaced posteriorly and rotated under the base.

In lateral view the vault shows a low profile, a well-developed supraorbital torus and a supratoral groove. The inion is high and appears to coincide with the opisthocranion leaving a marked occipital *planum* below, bounded above by an occipital crest that fuses with a strong supramastoid crest. In occipital view the vault profile is broad at its base so that the maximum breadth of the skull is low.

The face is of particular interest since it is the only skull from the lower layers in Java that is at all well preserved in facial morphology. Its most striking features are the breadth of the zygomatic region and the massive nature of the supraorbital tori and the prominence of the glabella.

The teeth are heavily worn but resemble those known from other pithecanthropine specimens. The canine is larger than those of modern man and the molars diminish in length from the first to the third molar (Sartono, 1971, 1972).

Dimensions   Sartono (1971)

Pith. VIII		I^1	I^2	Upper Teeth (Crown Dimensions) C	PM^1	PM^2	M^1	M^2	M^3
Left	l	—	—	—	—	8·2	—	—	—
side	b	—	—	—	—	10·1	—	—	—
Right	l	—	—	8·9	—	—	10·9	10·7	9·4
side	b	—	—	10·5	—	—	12·9	12·9	13·1

Morphology (4)   MANDIBLE

The 1941 specimen consists of part of the right side of the body of a massive hominid mandible that extends from the canine socket to the first molar tooth. Three large teeth are *in situ*, the first and second premolars and the first molar. The jaw is remarkable in size, being larger than any known example from modern man, equalled by few modern gorillas

Fig. 100 The Meganthropus II mandibular fragment
(1941)
Courtesy of Professor G. H. R. von Koenigswald

and only exceeded by *Gigantopithecus*. The inner aspect of the fragment bears genial tubercles for the attachment of the extrinsic tongue muscles, and shows part of the digastric impression. There is no simian shelf. The mental foramen is placed about midway between the upper and lower borders of the bone.

TEETH

The first premolar is bicuspid and asymmetrical in occlusal view, and has two well defined grooves on its buccal surface. The second premolar bears two mesial cusps and a talonid basin. The fused ridges joining the cusps separate the anterior and posterior foveae while the buccal grooves are feebly 298 represented. The first molar is a robust tooth but attrition has exposed the dentine leaving little of the fissural pattern. The occlusal surface is elongated mesiodistally. Despite the extensive wear it is probable that there were six cusps present.

Dimensions   *Weidenreich* (1954); *Marks* (1953)

MANDIBLES

Specimens	Body height at		$M_2/$	Body thickness at		$M_2/$
	Symphysis	Mental Foramen	M_3	Symphysis	Mental Foramen	M_3
1941 (cast)	47·0	48·0	45·0	25·5	28·0	26·3
1953	37·0	42·0	47·0	—	—	—

Affinities   1. There is little doubt that the Pithecanthropus II calvaria belonged to a hominid of the same type as that of the Trinil calotte; in consequence it was attributed to *Pithecanthropus* by von Koenigswald and Weidenreich (1939). This view was strongly contested by Dubois (1940) who alleged that this skull was really the remains of a Solo man (*Homo soloensis*) said to be synonymous with Wadjak man (*Homo wadjakensis*). Despite this controversy it was apparent that Pithecanthropus I and II were almost identical.

2. These remains were examined and described by Weiden-reich (1945a), and a tentative reconstruction of the skull was attempted. At first he believed that this calvaria was male and that the previously known specimens must be female. Later he abandoned this idea as he could not reconcile some of the features of the skull with this scheme of interpretation and assumed that it must belong to a different group, *Pithecan-thropus robustus*. Von Koenigswald (1950) could not accept this view. Subsequently it has become clear that Pithecan-thropus IV is closely allied to the pithecanthropines of the Far East, in particular those of Java, but may well have affini-ties to remains known from East Africa (Tobias and von Koenigswald, 1964).

3. The new specimen, Pithecanthropus VIII, shows unmistak-able similarities to the other pithecanthropines known from   299

Java although at present detailed comparative studies are not available. Its value will be not only that it increases the sample of hominids from Java but also that the presence of the face is unique.

4. Although the first fragments of this form were found by von Koenigswald in 1941 and 1950, because of wartime difficulties the 1941 fragment was first fully described from casts and named by Weidenreich (1954a). Weidenreich believed that it belonged to a hominid who was ancestral to *Pithecanthropus* and thus to modern man, denying any relationship with the australopithecines. This view was taken up vigorously by Robinson (1953) who suggested that *Meganthropus* is at least equivalent to *Paranthropus*.

Later, in a review of the classification of the australopithecines (Robinson, 1954), *Meganthropus palaeojavanicus* was renamed *Paranthropus palaeojavanicus*. This step was criticized by Remane (1954a and b), rejected by von Koenigswald (1954, 1957), but stoutly defended by Robinson (1955, 1962). Subsequently Tobias and von Koenigswald (1964) have compared the Javan *Meganthropus* jaw fragments and the Olduvai Hominid 7 mandible from Bed I Olduvai Gorge and drawn attention to a number of similarities. The third mandibular fragment, found by Marks in 1952 (Marks, 1953), has been considered by von Koenigswald (1968, 1973) who feels that it shows evidence of having been split by a crocodile; yet he agrees with Marks' view that it is a *Meganthropus*. While accepting the hominid nature of this material, von Koenigswald (1973) takes the view that *Meganthropus* is a 'terminal form', in contrast to the view expressed by Lovejoy (1970) who cannot detect a discontinuity between the *Meganthropus* specimens and the other *Homo erectus* specimens from similar beds.

In general terms it appears that von Koenigswald believes that the hominid material that is derived from the Putjangan beds (Djetis fauna) should be allocated to *Pithecanthropus modjokertensis* or to *Meganthropus palaeojavanicus* (von Koenigswald, 1973). A more widely held view would place most of the Sangiran hominids into *Homo erectus* (Mayr, 1950; Dobzhansky, 1944; Campbell 1964).

SANGIRAN REMAINS

British Museum Catalogue No.	Former Designation	Formation	Material	Finder and year of find
Sangiran 1a	*			
Sangiran 1b	Mandible B	Putjangan	Rt. mandible, P_4–M_3	Von Koenigswald, 1936
Sangiran 2	Pithecanthropus II	Kabuh	Calotte	Von Koenigswald, 1937
Sangiran 3	Pithecanthropus III	Kabuh	Parietals, occipital	Von Koenigswald, 1938
Sangiran 4	Pithecanthropus IV (Holotype: *P. robustus*)	Putjangan	Calvaria, maxilla, Lt. C–M^1, Rt. C–M^3	Von Koenigswald, 1938/39
Sangiran 5	(Holotype: *P. dubius*)	Putjangan	Rt. mandible, M_1 and M_2	Von Koenigswald, 1939
Sangiran 6	Meganthropus A (Mandible D) (Holotype: *M. palaeojavanicus*)	Putjangan	Mandible, Rt. P_3–M_1,**	Von Koenigswald, 1941
Sangiran 7a	—	Putjangan	Isolated teeth	— 1937–41
Sangiran 7b	—	Kabuh	Isolated teeth	— 1937–41
Sangiran 8	Meganthropus B	Kabuh	Mandible, Rt. M_3	Marks, 1952
Sangiran 9	Mandible C	Putjangan	Rt. mandible, C, P_3 and P_4, M_2 and M_3	Sartono, 1960
Sangiran 10	Pithecanthropus VI***	Kabuh	Calotte, Lt. zygoma	Jacob, 1963
Sangiran 11	—	Kabuh	Lt. M^3, Rt, I_1	Jacob, 1963
Sangiran 12	Pithecanthropus VII	Kabuh	Calotte	Sartono, 1965
Sangiran 13b	—	Kabuh	Calotte fragments	Jacob, 1965
Sangiran 14	—	Kabuh	Cranial fragments	Jacob, 1966
Sangiran 15a	—	Kabuh	Lt. maxilla, P^3 and P^4	Sartono, 1968
Sangiran 15b	—	Kabuh	Lt. maxilla, P^3, roots P^4	Jacob, 1969
Sangiran 16	—	Kabuh	Rt. M^2, Lt. P^3 (germ)	Jacob, 1969/70
Sangiran 17	Pithecanthropus VIII	Kabuh	Cranium, Rt. C, M^{1-3}, Lt. P^3	Sartono, 1969
Sangiran 18a	—	Kabuh	Calotte fragments	Jacob, 1970
Sangiran 19	—	Kabuh	Occipital	Jacob, 1970
Sangiran 20	—	Kabuh	Calvarial fragments	Jacob, 1970
Sangiran 21	Mandible E	Kabuh	Mandible, M^3	Sartono, 1973
Sangiran 22	F	Putjangan	Mandible, Lt. I_2–M_1, Rt. P_4–M_2 + Lt. M_{2-3}, Rt. M_3	Sartono, 1974
Sangiran 23 ****	—	Kabuh	Endocast	Jacob, 1975

* B.M. Cat. No. 1a may be part of a maxilla of *Meganthropus* not yet described (Pers. Comm. J.H.R. von K.).
** B.M. Cat. No. 6 includes 'Lt. M_2 and M_3' but these do not appear to be a part of this specimen.
*** Sangiran 10 is listed by Jacob (1966 and 1975) as 'P. V' or 'the sixth skull'.
**** A further specimen, a fragment of a mandible of *Meganthropus* with Lt. M_2 and M_3 is mentioned by Tobias and von Koenigswald (1964).

The Sangiran Remains

Originals   Natur-Museum und Forschunge-Institut Senckenberg, Senckenberg-Anlage 25, 6 Frankfurt-am-Main 1, W. Germany.

Casts   University Museum, University of Pennsylvania, Philadelphia, Pennsylvania 19104, U.S.A.

References   SELENKA, M. L. and BLANCKENHORN, M. 1911
Die Pithecanthropus—Schichten auf Java. Leipzig: Verlag von Wilhelm Engelmann.
KOENIGSWALD, G. H. R. VON 1938
Ein neuer Pithecanthropus-Schädel. *Proc. Acad. Sci. Amst. 41,* 185-192.
KOENIGSWALD, G. H. R. VON and WEIDENREICH, F. 1939
The relationship between Pithecanthropus and Sinanthropus. *Narute 144,* 926-929.
DUBOIS, E. 1940
The fossil human remains discovered in Java by Dr G. H. R. von Koenigswald and attributed by him to Pithecanthropus erectus, in reality remains of Homo wadjakensis (syn. Homo soloensis). *Proc. Acad. Sci. Amst. 43,* 494-496, 842-851, 1268-1275.
KOENIGSWALD, G. H. R. VON 1942
The South-African man-apes and Pithecanthropus. Washington: Carnegie Inst. Publ. No. 530, 205-222.
DOBZHANSKY, T. 1944
On species and races of living and fossil men. *Am. J. Phys. Anthrop. 2,* 251-265.
WEIDENREICH, F. 1945a
Giant early-man from Java and South China. *Anthrop. Pap. Am. Mus. Nat. Hist. 40,* 1-134.
WEIDENREICH, F. 1945b
The puzzle of Pithecanthropus. Science and scientists in the Netherlands Indies. New York: Board for the Netherlands Indies, Surinam and Curaçao.
KOENIGSWALD, G. H. R. VON 1949
The discovery of early man in Java and Southern China. In *Early man in the Far East.* Ed. W. W. Howells. Philadelphia. *Stud. Phys. Anthrop. 1,* 83-98.
KOENIGSWALD, G. H. R. VON 1950
Fossil hominids from the Lower Pleistocene of Java. *Proc. Ninth Internat. Geol. Congr., London 1948, Sect. 9,* 59-61.
MAYR, E. 1950
Taxonomic categories in fossil hominids. *Cold Spring Harbour Symposia on Quantitative Biology 15,* 109-118.

The Sangiran Remains

MARKS, P. 1953
Preliminary note on the discovery of a new jaw of Meganthropus, von Koenigswald, in the Lower Middle Pleistocene of Sangiran, Central Java. *Indones. J. Nat. Sci. 109*, 26-33.

ROBINSON, J. T. 1953
Meganthropus, australopithecines and hominids. *Am. J. Phys. Anthrop. 11*, 1-38.

KOENIGSWALD, G. H. R. VON 1954
The Australopithecinae and Pithecanthropus III. *Proc. Acad. Sci. Amst. 57*, 85-91.

REMANE, A. 1954a
Structure and relationships of Meganthropus africanus. *Am. J. Phys. Anthrop. 12*, 123-126.

REMANE, A. 1954b
Methodische, probleme der Hominiden—Phylogenie II *Z. Morph. Anthr. 46*, 225-268.

ROBINSON, J. T. 1954
The genera and species of the Australopithecinae. *Am. J. Phys. Anthrop. 12*, 181-200.

ROBINSON, J. T. 1955
Further remarks on the relationship between Meganthropus and australopithecines. *Am. J. Phys. Anthrop. 13*, 429-446.

KOENIGSWALD, G. H. R. VON 1957
Meganthropus and the Australopithecinae. *Proc. Third Pan-African Cong. Prehist., Livingstone, 1955*, 158-160. Ed. J. D. Clark. London: Chatto and Windus.

PIVETEAU, J. 1957
Traité de Paléontologie, VII. Paris: Masson et Cie.

SARTONO, S. 1961
Notes on a new find of a Pithecanthropus mandible. *Publikasi Teknik Seri Paleontologi. no. 2.*

HOOIJER, D. A. 1962
The Middle Pleistocene fauna of Java. In *Evolution und Hominisation*, 108-111. Ed. J. Kurth. Stuttgart: Gustav Fischer Verlag.

KURTÉN, B. 1962
The relative ages of the australopithecines of Transvaal and the pithecanthropines of Java. In *Evolution und Hominisation*, 74-80. Ed. G. Kurth. Stuttgart: Gustav Fischer Verlag.

ROBINSON, J. T. 1962
The origin and adaptive radiation of the australopithecines. *Ibid.*, 120-140.

CAMPBELL, B. 1964
Quantitative taxonomy and human evolution. In *Classification and human evolution*, 50-74. Ed. S. L. Washburn. London: Methuen and Co. Ltd.

KOENIGSWALD, G. H. R. VON 1964
Potassium-argon dates and early man: Trinil. *Sixth Conf. Int. Ass. Quatern. Res. Warsaw 1961*, 325-327.

The Sangiran Remains

TOBIAS, P. V. and KOENIGSWALD, G. H. R. VON 1964
A comparison between the Olduvai hominines and those of Java and some implications for hominid phylogeny. *Nature 204*, 515-518.

JACOB, T. 1966
The sixth skull cap of Pithecanthropus erectus. *Am. J. Phys. Anthrop. 25*, 243-260.

KOENIGSWALD, G. H. R. VON 1968
Observations upon two Pithecanthropus mandibles from Sangiran Central Java. *Prcc. Acad. Sci. Amst. B. 71*, 99-107.

SARTONO, S. 1968
Early man in Java: Pithecanthropus skull VII, a male specimen of Pithecanthropus erectus (I). *Proc. Acad. Sci. Amst. B. 71*, 396-422.

LOVEJOY, C. O. 1970
The taxonomic status of the 'Meganthropus' mandibular fragments from the Djetis beds of Java. *Man 5*, 228-236.

SARTONO, S. 1971
Observations on a new skull of Pithecanthropus erectus (Pithecanthropus VIII) from Sangiran, Central Java. *Proc. Acad. Sci. Amst. B. 74*, 185-194.

SARTONO, S. 1972
Discovery of another hominid skull at Sangiran, Central Java. *Curr. Anthrop. 13*, 124-126.

JACOB, T. 1972
The absolute date of the Djetis beds at Modjokerto. *Antiquity 46*, 148.

KOENIGSWALD, G. H. R. VON 1973
The oldest hominid fossils from Asia and their relation to human evolution. *Proc. of symposium L'origine dell'uomo, Rome 1971*. Quaderno. N. 182, Accademia Nazionale dei Lincei.

JACOB, T. 1975
The Pithecanthropines of Indonesia. *Bull. et Mem. de la Soc. d'Anthrop. de Paris t.2 série XIII*, 243-256.

OAKLEY, K., CAMPBELL, B., MOLLESON, T. 1975
Catalogue of fossil hominids. Part III: Americas, Asia, Australasia, 108-113. London: Trustees of the British Museum (Natural History).

The Ngandong Remains

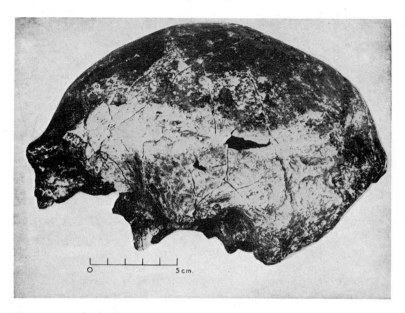

Fig. 101   Solo skull number 6
Left lateral view
Courtesy of Professor J. S. Weiner

Synonyms *Homo (Javanthropus) soloensis* (Oppenoorth, 1932a); *Homo*
and other names *soloensis* (Oppenoorth, 1932b); *Homo primigenius asiaticus*
(Weidenreich, 1933); *Homo neanderthalensis soloensis* (von
Koenigswald, 1934); *Homo sapiens soloensis* (Campbell, 1964)
Solo man; Ngandong man

Site Ngandong, six miles north of Ngawi, Central Java, Indonesia.

Found by C. ter Haar, 1931–1933; G. H. R. von Koenigswald, 1933.

Geology The valley of the river Solo, north of Ngawi, has three
gravel terraces at two metres, seven metres and 20 metres,
where the river has cut through the previous fluviatile deposits.
It was in the high 20-metre terrace, above the Kabuh beds,
that the Solo finds were uncovered.

305

The Ngandong Remains

Associated finds   A few small stone implements were found with the bones as well as some stone balls, but too few indisputable artefacts were recovered to constitute an industry. Several rayfish spines and deer antlers were found, which may have been used as spearpoints or pick-axes. A large quantity of mammalian bones were associated with the hominid remains, mostly belonging to an axis deer (*Cervus javanicus*) or banteng cattle. Other forms included pigs (*Sus terhaari, Sus macrognathus*), rhinoceros (*Rhinoceros sondaicus*), hippopotamus (*Hexaprotodon*

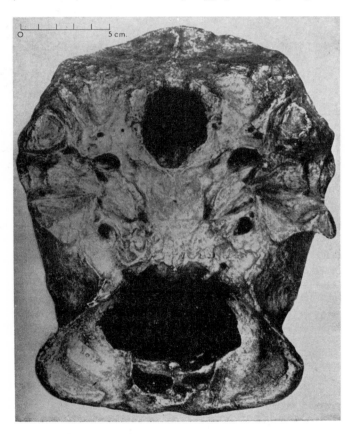

Fig. 102   Solo skull number 11
Basal view
Courtesy of Professor J. S. Weiner

ngandongensis) and primitive elephant (*Stegodon sp.*). The diagnostic fossils for the Ngandong fauna are *Cervus javanicus* and *Sus terhaari*.

Dating   Assessment of the stratigraphy and fauna suggest that the dating of the Ngandong deposits is Upper Pleistocene (von Koenigswald, 1949).

Morphology   CALVARIAE

In all eleven calvariae and two tibiae were unearthed at this site. Seven were regarded as being adult on grounds of sutural fusion (Weidenreich, 1951) and, of these, two were believed to be male, two female and the remainder of indeterminate sex.

The calvariae are all thick, several showing signs of injury during life. In profile they all possess the same general form which, in combination with similar dimensions, suggests that they represent a homogeneous population. Particular features of the profile are supra-orbital ridges separated by a central depression, sloping foreheads and strongly marked nuchal crests. The glenoid fossae are deep, and the articular eminences and mastoid processes pronounced. The base of No. 11 is complete apart from its anterior portion.

TIBIAE

Tibia A is broken at both ends, lacking articular surfaces, whereas Tibia B is nearly complete. Both bones are straight and appear modern in form.

Dimensions   CALVARIAE

Weidenreich (1951); Von Koenigswald (1958).
Max. Lengths 191–221   Max. Breadths 146–159
Cranial Indices 65·2–75·2   Cranial Capacities 1035–1255 cc
(Based on the six best-preserved skulls)

TIBIAE	*Length*
Tibia A	300 (broken)
Tibia B	365

★   ★   ★

CALVARIAE

Singer (1958)
Measurements taken on the original material.
Max. Lengths 192·5–220·3   Max. Breadths *c.* 144–*c.* 155
Cranial Indices 66·8–76·5

Cranial Lengths, Breadths and Length/Breadth Indices

Skull No.	1	5	6	9	10	11
Length	196·0	220·3	192·5	c. 201·0	202·6	200·0
Breadth	c. 148·0	c. 147·0	c. 144·0	c. 150·0	c. 155·0	c. 144·0
Cranial Index*	75·6	66·8	75·3	74·6	76·5	72·0

* Calculated from Singer's figures

Affinities The calvariae reported by Oppenoorth (1932a) were originally assigned by him to the genus *Homo* and placed in a sub-genus *Javanthropus*. The name was dropped in subsequent publications (Oppenoorth, 1932b, 1937) and the name *Homo soloensis* was proposed. However, Weidenreich (1933) suggested the name *Homo primigenus asiaticus* as part of a wider scheme of hominid classification. Vallois (1935) criticized the creation of a sub-genus for Solo man and suggested that these people were simply a local variety of Neanderthal man. Soon after this there followed a protracted controversy between Dubois and von Koenigswald regarding the relationships of all the Javan finds; Dubois (1936) at first proclaimed the 'racial identity' of Solo man, Modjokerto man and Peking man, but later believed that Solo man was identical with Wadjak man and thus a form of *Homo sapiens* (Dubois, 1940).

However, von Koenigswald had indicated his belief in the Neanderthal affinities of this form by naming it *Homo neanderthalensis soloensis* (von Koenigswald, 1934).

Finally Weidenreich examined all the Solo material in considerable detail and, in an unfinished paper (Weidenreich, 1951), elected not to enter the discussion on nomenclature; he contented himself by stating that 'Ngandong man is not a true Neanderthal type but distinctly more primitive and very close to *Pithecanthropus* and *Sinanthropus*'. None the less, von Koenigswald has adhered to his view that Solo man is a primitive 'tropical Neanderthaler' (von Koenigswald, 1958). In a recent classification of the Hominidae it has been proposed to include Solo man as a sub-species of *Homo sapiens* (*Homo sapiens soloensis*) distinct from modern man (*H. sap. sapiens*), Neanderthal man (*H. sap. neanderthalensis*) and Rhodesian man (*H. sap. rhodesiensis*) (Campbell, 1964).

Originals G. H. R. von Koenigswald Collection, Palaeoanthropological Division, Senckenberg Museum, Frankfurt am Main.

308

The Ngandong Remains

Casts   Not available at present.

References OPPENOORTH, W. F. F. 1932a
Homo (Javanthropus) soloensis, een plistoceene Mensch von Java.
Wet. Meded. Dienst. Mijnb. Ned.-O.-Ind. 20, 49-75.
OPPENOORTH, W. F. F. 1932b
De vondst paleolithische menschelijke schedels op Java. *De Mijningin-genieur 5,* 106-116.
KOENIGSWALD, G. H. R. VON 1933
Ein neuer Urmensch aus dem Diluvium Javas. *Zbl. Miner., Geol. A und B Paläont.,* 29-42.
WEIDENREICH, F. 1933
Ueber pithekoide Merkmale bei Sinanthropus pekinensis u seine stammesgeschichtliche Beurteilung. *Z. f. Anat. u Entw. Gesch. 99,* 212-253.
KOENIGSWALD, G. H. R. VON 1934
Zur stratigraphie des javanischen Pleistocän. *De Ing. Ned. Ind. 1* 185-201.
VALLOIS, H. 1935
Le *Javanthropus. Anthropologie 45,* 71-84.
DUBOIS, E. 1936
Racial identity of *Homo soloensis,* Oppenoorth (including *Homo modjokertensis,* von Koenigswald) and *Sinanthropus pekinensis,* David-son Black. *Proc. Acad. Sci. Amst. 39,* 1180-1185.
OPPENOORTH, W. F. F. 1937
The place of *Homo soloensis* among fossil men. In *Early Man,* 349-360. Ed. G. G. MacCurdy. Philadelphia and New York: J. B. Lippincott.
DUBOIS, E. 1940
The fossil human remains discovered by Dr G. H. R. von Koenigs-wald and attributed by him to *Pithecanthropus erectus,* in reality remains of *Homo wadjakensis* (syn. *Homo soloensis). Proc. Acad. Sci. Amst. 43,* 494-496, 842-851, 1268-1275.
KOENIGSWALD, G. H. R. VON 1949
The discovery of early man in Java and Southern China. *Stud. Phys. Anthrop. 1,* 83-98.
WEIDENREICH, F. 1951
Morphology of Solo man. *Anthrop. Pap. Amer. Mus. 43,* 205-290.
KOENIGSWALD, G. H. R. VON 1958
Der Solo-Mensch von Java; ein Tropische Neanderthaler. In *Hundert Jahre Neanderthaler,* 21-26. Ed. G. H. R. von Koenigswald. Utrecht: Kemink en Zoon.
SINGER, R. 1958
Ibid., p. 22.
CAMPBELL, B. 1964
Quantitative taxonomy and human evolution. In *Classification and human evolution,* 50-74. Ed. S. L. Washburn. London: Methuen and Co. Ltd.

The Wadjak Remains

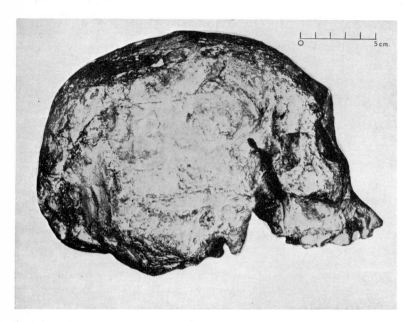

Fig. 103   Wadjak I skull
Right lateral view
Courtesy of Professor J. S. Weiner

Synonyms and other names	*Homo wadjakensis* (Dubois, 1921); *Homo sapiens wadjakensis* (Pinkley, 1936) Wadjak man
Site	Near Tulungagung, Central Java, Indonesia.
Found by	B. D. Van Rietschoten, 1889.
Geology	The skulls were found cemented in a limestone breccia terrace near an ancient lake.
Associated finds	No implements or other artefacts were associated with the skulls and the fossil fauna does not differ significantly from that found in modern Java.
Dating	In view of the degree of mineralization of the skulls and the modern fauna, the earliest possible dating is probably late Pleistocene.

310

Morphology The remains comprise a nearly complete skull and a broken mandible (Wadjak I), with a nearly complete mandible and a broken skull (Wadjak II). In addition there were some isolated teeth, and fragmentary bones, including a talus and two upper ends of tibiae.

SKULLS
The skulls are heavily mineralized and dolichocephalic. Wadjak I, apparently not fully prepared, was considered to be female by Dubois, whilst Wadjak II was believed to be male. Both skulls are large with well filled brain cases. The vault of Wadjak I is keeled with prominent superciliary ridges and a receding forehead; the occiput is protuberant, tending towards the formation of a bun.
The facial skeleton shows depression of the root of the nose, flattening of the nasal bones and a low position of the orbits. The maxillae present a marked degree of alveolar prognathism.

MANDIBLES
The mandibles are large and heavily built, the Wadjak II specimen bearing a definite chin.

TEETH
The Wadjak teeth are large but fall within the size range of modern Australian aboriginals. The upper molar size sequence is modern, $M_1 > M_2 > M_3$. The bite is edge to edge, there being neither overbite nor overjet in occlusion.

Molar Cusp Pattern
(*Lower Molars*)

	M_1	M_2
Wadjak I	Y5	+4
Wadjak II	Y5	+4

Dimensions *Dubois* (1922)
WADJAK I SKULL
Max. Length 200   Max. Breadth 145
Cranial Index 72·5 (Dolichocephalic)
Cranial Capacity 1,550 cc
WADJAK II SKULL
Cranial Capacity 1,650 cc

The Wadjak Remains

Affinities In the original description, published many years after the discovery of the material, Dubois (1922) drew attention to the Australoid features of Wadjak man, but elected to define a new species *Homo wadjakensis*. In a subsequent assessment, Pinkley (1936) suggested that this was unjustified and identified Wadjak man with *Homo sapiens*. This step has never been seriously contested and the Australoid affinities of Wadjak man have been widely accepted.

Originals Collection Dubois, Rijksmuseum von Natuurlijke Historie, Leiden, Netherlands.

Casts The University Museum, University of Pennsylvania, Philadelphia 4, Pennsylvania, U.S.A.

References DUBOIS, E. 1922
The Proto-Australian fossil man of Wadjak, Java. *Proc. Acad. Sci. Amst. 23,* 1013-1051.
PINKLEY, G. 1936
The significance of Wadjak man, a fossil *Homo sapiens* from Java. *Peking nat. Hist. Bull. 10,* 183-200.

The Peking and Lantian Remains

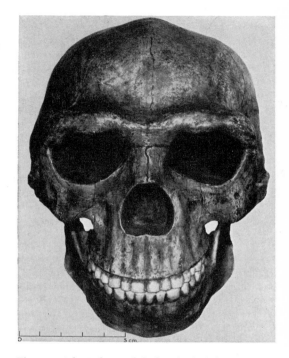

Fig. 104 The Peking adult female skull (cast)
Restored by F. Weidenreich
Frontal view
Courtesy of the Trustees of the British Museum
(*Nat. Hist.*)

Synonyms *Homo sp.* (Zdansky, 1927); *Sinanthropus pekinensis* (Black,
and other names 1927); *Pithecanthropus pekinensis* (Boule and Vallois, 1946);
Pithecanthropus sinensis (Piveteau, 1957); *Homo erectus pekinensis*
(Weidenreich, 1940; Campbell, 1964); ?*Sinanthropus lanti-*
anensis (Woo, 1964)
Pekin man; Peking man; Lantian man; Lantian 1 and 2
(Oakley *et al.*, 1975)

313

The Peking and Lantian Remains

Site
(i) *Skulls, teeth and postcranial bones.* The Choukoutien Lower Cave, near the village of Choukoutien, 25 miles south-west of Peking.
(ii) *Mandible* (Lantian 1). Chenchiawo, Lantian District, Shensi Province, north-west China.
(iii) *Skull* (Lantian 2). Foothills of the northern slope of the Chinling mountains, Kungwangling Hill (Gongwangling Hill), Lantian District, People's Republic of China.

Found by
J. G. Andersson, 1921; O. Zdansky, 1923; B. Bohlin, 1927; W. C. Pei, and the Cenozoic Research Laboratory, 1928–1937; Institute of Vertebral Palaeontology Team, 1949–1959; J. K. Woo 1959, 1963 and 1964; Institute of Vertebrate Palaeontology and Palaeoanthropology Team, 1966.

Geology
Choukoutien: The hills near Choukoutien are formed from Ordovician limestone which has been undermined and eroded by percolating ground waters producing caves and fissures. At Locality 1 a huge cavern roof has collapsed on top of the cave-filling, which is made of red clays and fallen rocks consolidated into a calcareous breccia. The cliff face at the principal site is 150 feet deep and was divided by Davidson Black into 15 sections each of ten feet, lettered A–O from above downwards. It was in this cave-filling, at various levels, that much of the material was found.

Lantian: The geology of this area has not been published in detail as yet. It seems, however, that the Chenchiawo mandible was found in the base of a layer of reddish clay 30 metres thick beneath which there is a layer of gravel one metre thick. The Lantian skull from Kungwangling was found in a block of 'fossil-bearing deposits' sent to the laboratory in Peking.

Associated finds
Choukoutien: The tools found at Locality 1 belong to a crude 'Chopper-tool' industry, and were made from imported coarse-grained quartz and greenstone. They are in the form of a few cores and numerous flakes which were probably utilized. The remains of an extensive mammalian fauna was recovered with Peking man. Amongst the forms recognized were some insectivores, bats and lagomorphs, numerous rodents, some

314

small and large carnivores, a large deer (*Megaloceros pachyo-steus*) and rhinoceros. A faunal list has been given by Kahlke (1962). At this site also were found ash heaps and pieces of charcoal; the charcoal, although of no use for radio-carbon dating on account of its age, provides the first clear evidence of the use of fire by early man.

Lantian: At both the skull and mandible sites, numerous quartz chopping tools, cores and flakes were recovered (Chia, 1966; Dai, 1966). The mammalian fauna found with the mandible includes red dog (*Cuon cf. alpinus*), tiger (*Felis cf. tigris*), elephant (*Elephantidae*), boar (*Sus lydekkeri*) and sika deer (*Pseudaxis grayi*) (Chia, 1966). The mammalian fauna found with the skull is predominantly of woodland forms including the sabre-toothed tiger (*Megantereos*), cheetah (*Acinonyx*), lion (*Felis leo*), tapir (*Tapirus indicus*) and giant macaque (*Macacus robustus*); most of those forms are poorly represented, but some grassland forms are better preserved (Woo, 1966).

Dating *Choukoutien:* The usually accepted dating of the Choukoutien site, in view of the fauna, is Middle Pleistocene. By a combina-tion of pollen and faunal analysis, Kurtén (1959) suggested that the dating equivalent of the Choukoutien deposits should be sought in the European glaciations and not in an interglacial, in his view probably the Second Glaciation (Elster II, Mindel II or the Antepenultimate Glaciation). The age of this glaciation, according to Evernden, Curtis and Kistler (1958) using the potassium–argon method, is 370,000 B.P. (a figure which may be revised to *c.* 400,000 B.P.). This is an appropriate date for the Choukoutien deposits but is dependent upon Kurtén's correlation. In a recent assessment of the geology of the cave-filling (Huang, 1960), it has been suggested that the deposits were laid down over a long period as six successive gravel beds; the basal layer during the First Glaciation, the lower two layers during the First Interglacial, and the upper three layers during the Second Glaciation. This correlation is rejected by Kahlke (1962), as were a previous Cromerian correlation and a Mindel–Riss Interglacial correlation proposed on palaeontological grounds.

Lantian: The dating of the Lantian sites suggests, on strati-graphic grounds, that they are earlier than that of Choukout-ien and are possibly contemporaneous with the earlier Javan *Homo erectus* sites (Woo, 1966).

Morphology   *Choukoutien:* The hominid remains from Choukoutien, which were described in a remarkable series of monographs by Black and Weidenreich, consist of 14 calvariae and 11 mandibles in varying states of preservation, as well as numerous teeth and a few post-cranial bones.
A new mandible has been attributed to Peking man by Woo and Chao (1959).

CALVARIAE
The first calvaria to be found, now known as Skull III, was recovered from Locus E. It is well preserved but the face and base are missing. The braincase is characterized by a flattened but keeled frontal region with a pronounced supra-orbital visor, a well marked post-orbital constriction and a prominent occipital torus. The mandibular fossae are deep and narrow. In frontal view the side walls of the cranium slope inwards towards the apex so that the maximum width lies in the region of the temporal bone above the small mastoid process—a feature not found in modern man. This particular skull was described by Black (1931) as being adolescent; later Weidenreich (1943) considered it to be juvenile. The principal morphological features of the remaining calvariae are similar to those of Skull III but tend to be more coarsely represented.

MANDIBLES
Three of the best-preserved mandibles have been recon-structed by Weidenreich (1936). These show the recession of the symphysial region, the narrow but rounded dental arcade, thickening of the body of the mandible inside the alveolar margin (mandibular torus), the presence of genial tubercles for the tongue muscles and an unusually large bicondylar breadth. A point of detail is the high incidence of multiple mental foramina (cf. Heidelberg jaw). The new mandible (1959) has a narrow alveolar arch, a moderate mandibular torus and four mental foramina on the right side.   316

TEETH

In total 147 teeth were examined by Weidenreich (1937), 83 socketed, 64 isolated. Out of the 132 (52 upper, 82 lower) permanent teeth present, every tooth was represented but only lower deciduous teeth are known. This collection is believed to come from 32 individuals, 20 of whom were adolescent or adult and 12 children. Five more teeth have been reported, but not described, by Woo (1960).

The most striking feature of the teeth is their variability in size, the range of which has permitted their division into two main groups believed to represent males and females. In general the teeth are robust and characteristically wrinkled. The upper incisors are shovel-shaped and frequently have a well developed basal tubercle with finger-like processes directed toward the free margin of the tooth. The upper canines are large and project beyond the occlusal line whilst the lower canines are smaller and tend to form cutting edges, thus resembling incisors. There is no trace of a diastema and the premolars are non-sectorial. In almost all of the canines, premolars and molars there is a cingulum which in the case of the molars has traces of stylid cusps. The cusp pattern is basically dryopithecine with a tendency towards transformation into the 'plus' pattern by reduction of the metaconid. The first and second permanent molars are of approximately the same size, but the third molar tends to be smaller than either. The permanent molars, premolars and milk molars display a degree of enlargement of the pulp cavity (taurodontism).

The eruption order of the teeth of Peking man differs from that of modern man in that the second permanent molar arises before both premolars and canine.

POSTCRANIAL BONES

The limb remains from Choukoutien include seven fragments of femoral shaft, none with an articular surface. The femora are unusual in the thickness of the cortical bone, the platymeria of the shaft, the distal positions of the narrowest point of the shaft and the convexity of the medial border of the bone (Weidenreich, 1941). The upper limb remains consist of two fragments of humerus, a lunate bone and a fragment said to be 317 part of a clavicle. (The examination of a cast of the 'clavicular' fragment has raised doubts as to its correct identification.) The humeral fragments have no articular surface, but are thick walled and one has a strong deltoid impression. The lunate belongs to the right side and is small by comparison with those of modern man. The bone is somewhat eroded having lost part of its dorsal surface, part of the semi-lunar facet for the scaphoid and the apex of the ridge joining the radial and triquetral surfaces.

Woo (1960) reported new limb material—'two fragments of humerus and tibia'. The tibial fragment is said to have an even smaller cavity than the femora. No further details of the post-cranial bones have been published yet.

Lantian: In 1963 a well-preserved mandible was recovered from Chenchiawo, Lantian in Shensi province. Later in 1964 a tooth was recovered from Kungwangling hill at a height of 80 metres. Subsequently from a block recovered nearby a

Fig. 105 The Lantian mandible (Lantian 1)
Occlusal view
Courtesy of J. K. Woo

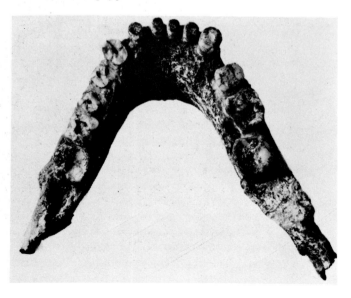

skull was found. It consisted of a calotte with a right temporal bone, parts of the orbits, parts of the nasal bones, parts of the left maxilla, and the right maxilla bearing two teeth. They appear to belong to one individual. The bones are somewhat distorted but the skull has the same general character as other *Homo erectus* skulls from Java and Choukoutien. In detail it appears to resemble earlier Javan specimens in terms of post-orbital constriction, lowness of the forehead, the size of the supraorbital region and the thickness of the vault bones.

Fig. 106   The Lantian skull (Lantian 2)
Internal view
Courtesy of J. K. Woo

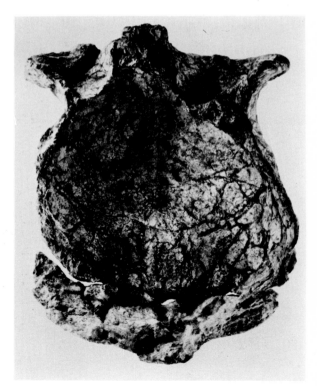

Dimensions CALVARIAE
Locus E. Skull III (Black, 1931)
Length (glabella/occipital) 187·6   Max. Parietal Breadth 133·0
Cranial Index 73·5
   (Dolichocephalic)
Cranial Capacity 915 cc
   (Weidenreich)

CRANIA
Weidenreich (1943)
Max. Length range 188–199
Max. Breadth range 137·2–143
Cranial Index range 71·4–72·6 (Dolichocephalic)
Cranial Capacity (5 skulls) range 915 cc–1,225 cc

Lantian 2. Woo (1966)

Cranial capacity 778 cc (Pearson's formula)
            775–783 cc (Tobias' partial endocast
                  method, Tobias, 1964).

SKULL THICKNESS
Glabella 24·0
Bregma 16·0
Squamous temporal 11·5

MANDIBLES G1♂ and H1♀
Reconstructed by Weidenreich (1936)

					At Mental Foramen	
		Bicondylar	*Symphysial*	*Ramus*		*Thick-*
No.	*Length*	*Breadth*	*Height*	*Height*	*Height*	*ness*
G1 ♂	103·0*	146·4*	40·0*	66·7	34·0	16·4
H1 ♀	94·0*	101·8*	31·5	59·0	26·0	15·4

* Restored.

The differences between these measurements were taken as evidence of sexual dimorphism.

320

The Peking and Lantian Remains

MANDIBLE

Sinanthropus lantianensis (Woo, 1964, 1966)

No.	Height at Mental Foramen	Symphysial Height	Thickness at M1
P.A.102	27·0	35·0	16·0

Lantian 1

TEETH
Weidenreich (1937)

Upper Teeth	Permanent Dentition (Crown Dimension Ranges)	
	Length	Breadth
Central Incisors (I1)	9·8–10·8	7·5– 8·1
Lateral Incisors (I2)	8·2– 8·3	8·0– 8·2
Canines (C)	8·5–10·5	9·8–10·6
First Premolars (PM1)	7·4– 9·2	10·5–12·8
Second Premolars (PM2)	7·2– 8·9	10·3–12·5
First Molars (M1)	10·0–13·1	11·7–13·7
Second Molars (M2)	10·2–12·2	12·2–13·4
Third Molars (M3)	8·7–10·4	10·4–12·5

Lower Teeth	Length	Breadth
Central Incisors (I1)	6·0– 6·8	5·8– 6·8
Lateral Incisors (I2)	6·3– 7·2	6·4– 7·3
Canines (C)	8·1– 9·0	8·2–10·4
First Premolars (PM1)	7·9– 9·8	9·1–10·8
Second Premolars (PM2)	8·2– 9·2	8·0–11·1
First Molars (M1)	9·9–13·6	10·1–12·8
Second Molars (M2)	11·3–13·1	11·1–12·9
Third Molars (M3)	10·0–13·8	10·0–12·4

Lower Teeth	Deciduous Dentition (Crown Dimensions)	
	Length	Height
Central Incisors (DI1)	4·3	3·6
Lateral Incisors (DI2)	—	—
Canines (DC)	6·1–6·2	5·2– 5·3
First Molars (DM1)	9·8	8·4–10·1

POST-CRANIAL BONES
Weidenreich (1941)
Lunate: Length (prox./dist. diam.) 14·5
     Breadth (radio/ulnar diam.) 14·4
     Height (dorso/volar) 16·5
Clavicle: Length (145)
     Circumference 34 (Mid-point)
Humerus: Angle of Torsion 137°
     Length (324)
Femora: Lengths (400–407)
     Mid-shaft Widths 29·2–29·7
     Mid-shaft Depths 22·8–27·1
( ) Estimated measurement

Affinities Initially the name *Sinanthropus pekinensis*, conferred by Davidson Black (1927), referred solely to the hominid molar tooth upon which the genus was founded. It was not long before Black's bold step was apparently justified by the subsequent finds, but soon afterwards the resemblances between *Pithecanthropus* and *Sinanthropus* began to become clear (Boule, 1929). Later, when the Choukoutien material had been fully studied by Weidenreich, he took the view that the differences between *Pithecanthropus* and *Sinanthropus* were of a racial character only. It has been suggested that Peking man, in company with several other Middle Pleistocene hominids, should be classified under the name of *Homo erectus* (Mayr, 1950) and given the geographic subspecific designation *pekinensis*. This proposal has been incorporated in a recent classification (Campbell, 1964). The Lantian material has been allocated to a new species of the genus '*Sinanthropus*', '*Sinanthropus lantianensis*' (Woo, 1965). It seems likely, however, that many anthropologists will regard the new material as further evidence of *Homo erectus* from the Far East.

Originals All of the early material, except one lower premolar and one upper molar, was lost during the 1939–45 war. Fortunately most of the bones had been cast and studied intensively by Weidenreich. The two teeth are at the University of Uppsala, Sweden. The newer material is kept at The Institute of Vertebrate Palaeontology and Palaeoanthropology, Academia Sinica, Peking, Peoples' Republic of China.

The Peking and Lantian Remains

Casts   Choukoutien

The University Museum, University of Pennsylvania, Philadelphia 4, Pennsylvania, U.S.A.

Lantian Man

Casts not available as yet.

References   ZDANSKY, O. 1927
Preliminary notice on two teeth of a hominid from a cave in Chihli (China). *Bull. geol. Soc. China.* 5, 281-284.
BLACK, D. 1927
On a lower molar hominid tooth from the Chou Kou Tien deposit. *Palaeont. sinica*, Ser. D, 7, 1-29.
BOULE, M. 1929
Le *Sinanthropus*. *Anthropologie* 39, 455-460.
BLACK, D. 1931
On an adolescent skull of *Sinanthropus pekinensis* in comparison with an adult skull of the same species and with other hominid skulls, recent and fossil. *Palaeont. sinica*, Ser. D, 7, II, 1-145.
WEIDENREICH, F. 1936
The mandibles of *Sinanthropus pekinensis*: a comparative study. *Palaeont. sinica*, Ser. D, 7, III, 1-163.
WEIDENREICH, F. 1937
The dentition of *Sinanthropus pekinensis*: a comparative odontography of the hominids. *Palaeont. sinica*, New Ser. D, 1, 1-180, 1-121 (plates).
WEIDENREICH, F. 1940
Some problems dealing with ancient man. *Amer. Anthrop.* 42, 375-383.
WEIDENREICH, F. 1941
The extremity bones of *Sinanthropus pekinensis*. *Palaeont. sinica*, New Ser. D, 5, 1-150.
WEIDENREICH, F. 1943
The skull of *Sinanthropus pekinensis*; a comparative study on a primitive hominid skull. *Palaeont. sinica*, New Ser. D, 10, 1-291.
BOULE, M., and VALLOIS, H. V. 1946
Les hommes fossiles. 3rd Ed., 122. Paris: Masson et Cie.
MAYR, E. 1950
Taxonomic categories in fossil hominids. *Cold Spring Harbour Symposia on Quantitative Biology* 15, 109-118.
PIVETEAU, J. 1957
Traité de Paléontologie VII, 384. Paris: Masson et Cie.
KURTÉN, B. 1959
New evidence on the age of Pekin man. *Vertebrata Palasiatica* 3, 173-175.

WOO, J. K., and CHAO, T. K. 1959
New discovery of *Sinanthropus* mandible from Choukoutien. *Vertebrata Palasiatica 3*, 169-172.

HUANG, W. P. 1960
Restudy of the Choukoutien *Sinanthropus* deposits. *Vertebrata Palasiatica 4*, 45-46.

HUANG, W. P. 1960
On the age of basal gravel of Choukoutien *Sinanthropus* site and of the 'Upper gravel' and 'Lower gravel' of the Choukoutien region. *Ibid.*, *4*, 47-48.

WOO, J. K. 1960
The unbalanced development of the physical features of *Sinanthropus pekinensis* and its interpretation. *Vertebrata Palasiatica 4*, 17-26.

KAHLKE, H. D. VON 1962
Zur relativen Chronologie ostasiatischer Mittelpleistozän-Faunen und Hominoidea-Funde. In *Evolution und Hominisation*, 84-107. Ed. G. Kurth. Stuttgart: Gustav Fischer Verlag.

ANON 1963
Lantian jaw. *Ill. Lond. News 243*, 742.

WOO, JU KANG 1964
A newly discovered mandible of the Sinanthropus type—*Sinanthropus lantianensis*. *Scientia Sinica 13*, 891-811.

WOO, JU KANG 1964
Mandible of *Sinanthropus lantianensis*. *Curr. Anthrop. 5*, 98-99.

WOO, JU KANG 1965
Preliminary report on a skull of *Sinanthropus lantianensis*, of Lantian, Shensi. *Scienta Sinica 14*.

WOO, JU KANG 1966
The Skull of Lantian Man. *Curr. Anthrop. 7*, 83-86.

CHIA, L. P. 1966
In *Transactions of the field conference on the Cenzoic of Lantian, Shensi*, Peking, 151-154 (in Chinese).

DAI, E. J. 1966
The Palaeoliths found at Lantian Man locality of Gongwangling and its vicinity. *Vertebrata Palasiatica 10*, 30-34.

OAKLEY, K. P., CAMPBELL, B. G. and MOLLESON, T. I. 1975
Catalogue of fossil hominids Part III: Americas, Asia, Australasia. London: Trustees of the British Museum (Natural History).

Glossary

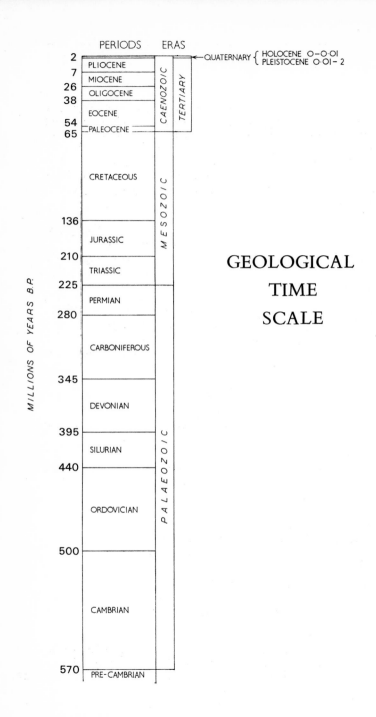

GEOLOGICAL

TIME

SCALE

MILLIONS OF YEARS B.P.

PERIODS	ERAS

QUATERNARY { HOLOCENE 0−0·01
PLEISTOCENE 0·01−2

2
PLIOCENE
7
MIOCENE
26
OLIGOCENE
38
EOCENE
54
PALEOCENE
65

CAENOZOIC

TERTIARY

CRETACEOUS

MESOZOIC

136
JURASSIC
210
TRIASSIC
225
PERMIAN
280
CARBONIFEROUS
345
DEVONIAN
395
SILURIAN
440
ORDOVICIAN
500
CAMBRIAN
570
PRE-CAMBRIAN

PALAEOZOIC

Glossary

ACHEULEAN
A stone tool culture characterized by distinctive pointed or almond-shaped hand-axes. Type site, St Acheul, Amiens (Somme), Northern France.

ALVEOLAR PROGNATHISM
Forward projection of the portions of the jaws that bear teeth.

AMINOACID RACEMIZATION METHOD
A dating method that depends upon the detection of residual aminoacids in fossil bone.

ANGLE OF THE CRANIAL BASE
The angle between the basi-occiput and the body of the sphenoid.

APPENDICULAR SKELETON
The bones of the limbs and the limb girdles.

ARTEFACTS (ARTIFACTS)
Man-made objects.

ARTESIAN WATER
Ground water contained under pressure.

ARTIODACTYLA
The zoological name given to an order of ungulates. The 'even-toed' ungulates, e.g. deer, antelopes, gazelle, buffalo, pigs, oxen, sheep and goats.

ASTERION
A point on the skull at which the lambdoid, parieto-mastoid and occipital sutures meet.

ATLANTHROPUS
The generic name given to a group of North African Middle Pleistocene hominids. After the Atlas range of mountains, North Africa.

AURIGNACIAN
A culture of stone, bone and antler which includes flint side and end scrapers, spear tips and blades. The bone implements include awls and spear points. The culture is associated with the Cro-Magnon people. Type site; Aurignac, Haute Garonne, France.

AUSTRALOPITHECINAE
The zoological sub-family which contains the fossil 'ape-men', 'man-apes' and 'near men'.

AUSTRALOPITHECINE
(*n.*) A member of the zoological sub-family Australopithecinae.
(*adj.*) Pertaining to the zoological sub-family Australopithecinae.

AUSTRALOPITHECUS
The generic name given to a group of South and East African Lower Pleistocene hominids. 'Southern ape.'

AXIAL SKELETON
The skull, vertebral column and thorax.

b
Breadth.

BASALT
Fine-grained extrusive igneous rock of dark colour, low in silica.

BENTONITIC CLAY
A clay formed from decomposed volcanic ash.

BICUSPID
(*n.*) A premolar tooth.
(*adj.*) Two-cusped.

BIPEDAL GAIT
Two-legged walking.

B.P.
Before present.

BRACHYCEPHALIC
Having a Cranial Index above 80. 'Broad-headed.'

BRECCIA
Sedimentary rock composed of angular fragments of derived material embedded in a finer cement.

327

Glossary

BREGMA
The point at which the coronal and sagittal sutures of the skull meet.

BURIN
An Upper Palaeolithic chisel-like stone tool suitable for engraving bone, wood, horn or soft stone.

CAENOZOIC (CENOZOIC)
See geological time scale.

CALCITE
The stable crystalline form of calcium carbonate at normal temperatures, hence the commonest mineral form of limestone.

CALCRETE
Desert soil cemented with calcium carbonate.

CALOTTE
The bones of the cranial vault.

CALVARIA
A skull which has lost the bones of the face including the mandible.

CAMBRIAN
See geological time scale.

CARABELLI'S CUSP
An accessory cusp on the lingual surface of the crown of an upper molar tooth.

CARABELLI'S PIT AND GROOVES
Features found on the lingual surface of the crown of an upper molar tooth.

CATARRHINE MONKEYS
A sub-group of the order Primates which includes monkeys from Africa and Asia. 'Old-world monkeys.'

CHÂTELPERRONIAN
An early phase of the Perigordian (*q.v.*).

CHERT
A siliceous rock found in limestone of which flint is an example.

CHIGNON
An occipital 'bun-like' protuberance of the skull characteristic of the Neanderthalers of the Fourth Glaciation.

CINGULUM
A collar-like ridge of enamel around the base of the crown of a tooth.

CLACTONIAN
A primitive flake culture found in Europe which includes concave scrapers, cores and some flakes with retouched edges. Type site, Clacton-on-Sea, Essex.

CONGLOMERATE
Sedimentary rock composed of rounded pebbles of older rocks embedded in a younger cement, e.g. puddingstone.

CRANIAL INDEX
$$\frac{\text{Max. Cranial Breadth}}{\text{Max. Cranial Length}} \times 100$$

CRANIUM
That part of the skull forming the brain-case.

CRASSIDENS
The specific name given to a group of South African hominids known from Swartkrans. 'Big-toothed.'

CYPHANTHROPUS
The generic name formerly given to Rhodesian man. 'Stooping-man.'

DENTAL CARIES
A pathological process, with destruction of tooth enamel and dentine, leading to infection and loss of the tooth.

DEVONIAN
See geological time scale.

DIABASE
An American term for dolorite. A medium-grained dark igneous rock, low in silica.

DIASTEMA
A gap between the teeth.

328

DIOPSIDE
A colourless or pale green mineral of calcium magnesium silicate.

DOLICHOCEPHALIC
Having a Cranial Index of less than 75. 'Long-headed.'

DOLOMITE
1. Mineral calcium magnesium carbonate.
2. Limestone containing more than 50% mineral dolomite.

DRIPSTONE
Crystalline calcium carbonate deposited from water in layers or strata.

DRYOPITHECINE
(*n.*) A member of the genus *Dryopithecus*.
(*adj.*) Pertaining to the genus *Dryopithecus*.

DRYOPITHECUS
The generic name given to a group of Middle Miocene apes. 'Oak-ape.'

ENAMEL WRINKLING
Secondary folding of the enamel of the occlusal surface of a tooth; consistently found in the molar teeth of the modern orang-utan (*Pongo*), also known in australopithecine and pithecanthropine teeth.

ENDOCAST
A cast of a cavity, displaying internal surface features in relief.

FAUNAL BREAK
A sudden change in the character of the fossil fauna encountered during excavation of successive layers of a deposit; possibly due to partial erosion and subsequent redeposition of later material containing a different fauna, or to faunal migration.

FAURESMITH
A Palaeolithic stone tool culture found in South Africa consisting of hand-axes and cleavers. Type site, Fauresmith, Orange Free State, Republic of South Africa.

FELSPAR (FELDSPAR)
A group of crystalline minerals consisting of silicates of aluminium with sodium, potassium, barium or calcium. Decomposition of felspars produces clays.

FEMORO-CONDYLAR ANGLE
The angle between the shaft of the femur and a line drawn tangentially to the articular surfaces of the femoral condyles.

FERRICRETE
Soil cemented with iron oxide.

FISSION TRACK METHOD
A radiometric dating method dependent upon the presence of a radioactive constituent in natural glasses.

FLUVIATILE DEPOSITS
Deposits produced by river action.

FRANKFURT PLANE
An agreed plane in which skulls may be oriented for comparative purposes. Arranged horizontally, it passes through the lower orbital margin and forms a tangent to the upper margin of the external auditory meatus.

HABILIS
The specific name given to a group of East African Lower Pleistocene hominids. 'Able, handy, vigorous, mentally skilful.'

HOLOCENE
See geological time scale.

HOMINID
(*n.*) A member of the zoological family Hominidae.
(*adj.*) Pertaining to the zoological family Hominidae.

329

Glossary

HOMINIDAE
The zoological family which includes fossil and modern man as well as the fossil 'ape-men', 'man-apes' and 'near-men'.

HOMININAE
The zoological sub-family which contains fossil and modern man.

HOMININE
(*n.*) A member of the zoological sub-family Homininae.
(*adj.*) Pertaining to the zoological sub-family Homininae.

HOMINOID
(*n.*) A member of the zoological super-family Hominoidea.
(*adj.*) Pertaining to the zoological super-family Hominoidea.

HOMINOIDEA
The zoological super-family which contains fossil apes, 'ape-men', 'man-apes', 'near-men' and men, as well as modern great apes and modern man.

HOMO
The generic name given to the group of hominids which contains fossil and modern man.

HYPSODONT
Having teeth with tall crowns, e.g. the horse.

ILMENITE
Mineral iron titanium oxide.

INION
A position on the skull marked by the external occipital protuberance.

INTERGLACIAL
A warm period between two major glaciations.

INTERPLUVIAL
A dry phase between two rainy periods.

INTERSTADIAL
A warm interval within a major glaciation.

JAVANTHROPUS
The sub-generic name formerly given to a group of hominids from Ngandong, Java. 'Java-man.'

KARSTIC CAVES
Caves formed in limestone by the action of water.

l
Length.

LACUSTRINE
Lacustrine deposits are laid down in relatively still water lakes.

LAMBDA
A point on the skull at which the sagittal and lambdoid sutures meet.

LEVALLOISIAN
A Palaeolithic flake tool culture produced by striking serviceable flakes from a prepared core; recognized in Europe, Asia and Africa. Type site, Levallois-Perret, Paris.

LIAS (LIASSIC)
A rock formation consisting of layers of limestone, marl and clay attributed to the Lower Jurassic period. ? Corruption of 'layers'.

LISSOIR
A polishing or rubbing tool.

LOAM
An iron-rich mixture of clay and silt.

LOESS
A fine-grained deposit of wind-blown material.

MAMELON
A small hillock or tuberosity; small elevations found along the free margin of a newly erupted incisor tooth.

MEGANTHROPUS
The generic name given to a group of early hominids from Java and East Africa. 'Big-man.'

330

Glossary

MESOCEPHALIC (MESATICEPHALIC)
Having a Cranial Index which lies
between 75-80. 'Middle-headed.'

MESOZOIC
See geological time scale.

MOUSTERIAN
A stone tool culture consisting of side
scrapers, disc cores and points widely
represented in Europe, North Africa
and Western Asia; frequently associated with Neanderthal remains. Type
site, Le Moustier, Peyzac, Dordogne,
France.

NEUROCRANIUM
That part of a skull that makes up the
brain case.

NUCHAL
Pertaining to the nape of the neck.

OBSIDIAN
Volcanic glass, capable of fine con-
choidal fracture producing sharp flakes.

OCCLUSAL SURFACE
The biting surface of a tooth.

OLDOWAN
An Early Palaeolithic stone tool culture
consisting of crudely made cutting,
scraping or chopping implements produced by flaking stones in two direc-
tions. Type site, Olduvai Gorge,
Tanzania, East Africa.

ORTHOGNATHOUS
Without forward projection of either
upper or lower jaw. 'Straight-jawed.'

OSTEOARTHRITIS
A degenerative disease of joints characterized by pain, swelling and deformation.

OSTEODONTOKERATIC
Pertaining to bone, tooth and horn.

OVERBITE
The degree of vertical overlap of the
incisor teeth in occlusion.

OVERJET
The degree of separation of the upper
and lower incisor teeth in the horizontal plane when the teeth are occluded.

PALAEOMAGNETISM
Residual magnetism detectable in rock.

PALAEOZOIC
See geological time scale.

PALYNOLOGY
The study of pollen

PARANTHROPUS
A generic name given to a group of
South African Pleistocene hominids.
'Beside or equal-to-man.'

PEARSON'S FORMULA
Female Cranial Capacity $= 0.000,375$
$\times$ Length $\times$ Breadth $\times$ Height $+ 296.4$.
Male Cranial Capacity $= 0.000,365$
$\times$ Length $\times$ Breadth $\times$ Height $- 359.34$.

PERIGORDIAN
An Upper Palaeolithic tool tradition.

PERISSODACTYLA
The zoological name given to an order
of ungulates. The 'odd-toed' ungulates,
e.g. equids, rhinoceroses and tapirs.

PITHECANTHROPINE
(*n.*) A member of the genus *Pithecanthropus*.
(*adj.*) Pertaining to a member of the
genus *Pithecanthropus*. Frequently used
colloquially when referring to hominids of the *Homo erectus* group.

PITHECANTHROPUS
The generic name given to a group of
Asian Middle Pleistocene hominids.
'Ape-man.'
(Members of this genus are commonly
included within the genus *Homo* by
modern taxonomists.)

PLATYMERIA
Anteroposterior flattening of the shafts
of femora.

331

Glossary

PLESIANTHROPUS
The generic name given to a group of South African Lower Pleistocene hominids. 'Near-man.'

PLUVIAL
A rainy period.

POLYPHYLETIC
Concerning, or derived from several ancestral forms.

PONGID
(*n.*) A member of the zoological family Pongidae.
(*adj.*) Pertaining to the zoological family Pongidae.

PONGIDAE
The zoological family which contains both fossil and modern great apes.

POTASSIUM/ARGON METHOD
A radiometric dating method.

PROGNATHOUS
Forward projection of the jaws.

PROMETHEUS
The specific name given to a group of Transvaal Lower Pleistocene hominids in the mistaken belief that their remains were associated with evidence of fire.

QUADRITUBERCULATE
Bearing four tubercles.

QUARTZITE
A siliceous metamorphic rock consisting of quartz grains, or minute quartz crystals, set in a quartz cement.

RACLOIRS
Mousterian side scrapers.

ROBUSTICITY INDEX
An index obtained by expressing the thickness of a bone in terms of its length.

SECTORIAL TEETH
Teeth having a cutting edge and a scissor-like action in occlusion.

SEPSIS
Bacterial infection with the formation of pus.

SESAMOID
A bone formed within a tendon.

SHOVELLED INCISORS
Incisor teeth that are scooped out on their lingual surfaces and having a variable degree of inrolling of their lateral borders.

SILCRETE
Soil cemented with silica.

SIMIAN SHELF
A buttress of bone that reinforces the symphysial region of the mandible in monkeys and apes.

SINANTHROPUS
The generic name formerly given to a group of Middle Pleistocene hominids found near Peking, Peoples' Republic of China. 'Chinese-man.'

SKULL
The bony skeleton of the head, including the lower jaw. The term is not always strictly applied.

STALACTITE
A conical or irregular deposit of calcite hanging from the roof of a cave; formed by precipitation of calcium carbonate from drops of lime-saturated water.

STRATIGRAPHY
A branch of geology concerned with the formation, constituents and sequence of stratified deposits.

STILLBAY
A Late Palaeolithic stone tool culture found in South Africa consisting of finely made blades and weapon heads. Type site, Stillbay, Cape Province, Republic of South Africa.

SUPERCILIARY RIDGES
Smoothly rounded ridges of bone found on the frontal region of the skull above the position of the eyebrows.

332

Glossary

SYLVIAN CREST
A ridge found on the internal surface of the parietal bone of some primitive hominid skulls; it occupies the Sylvian fissure of the brain.

SYMPATRIC
Two or more animal populations that occupy the same habitat but do not interbreed are termed sympatric.

SYMPHYSIAL ANGLE
(SYMPHYSEAL ANGLE)
The angle made by the principal dimension of the mandibular symphysis and the lower border of the body of the mandible.

TAURODONT
Teeth having enlarged pulp cavities. 'Bull-toothed.'

TAYACIAN
A rather poorly defined flake culture, allied to the Clactonian, found in Europe and the Near East. Type site, Tayac, Les Eyzies, Dordogne, France.

TEKTITES
Glassy objects of supposed extra-terrestrial origin.

TELANTHROPUS
The generic name given to a group of South African hominids known from Swartkrans. ?'Distant-man' or 'Perfect-man.'

THORIUM/URANIUM METHOD
A radiometric dating method.

TORUS
A smooth rounded protuberance.

TRAVERTINE
Almost pure calcium carbonate rock deposited around lime-rich springs and lakes.

TROPICAL SAVANNAH
Tropical grassland containing scattered trees, such as the Baobab in Africa.

TUFA
A calcareous deposit, usually spongy in texture formed near lime-rich springs and rivers.

TUFF
A consolidated deposit of volcanic ash ften laid down in water.

VILLAFRANCHIAN
1. A faunal assemblage containing new types of mammals such as *Elephas* (*Archidiskodon*), *Equus*, *Bos* (*Leptobos*) and *Camelus*, which appeared suddenly during the Lower Pleistocene. Type site, Villafrancha d'Asti, Italy.
2. Pertaining to the Lower Pleistocene.

WORMIAN BONES
Sutural bones formed from isolated centres of ossification between major components of the skull vault. Commonly found between the occipital and parietal bones.

'ZINJANTHROPUS'
The generic name given to an East African Lower Pleistocene hominid. 'East African man'.

INDEX

Index

Main references to fossil hominids are in **bold type**. References to charts, diagrams and tables are in *italics*

Acheuléen II industry, 117
Acheulian culture, 33, 66, 199
  final, 274
  Middle, 19, 47, 199
  Upper, 84
  industry, 134
Acinonyx, 315
acoustic meatus length, 202
Aepyceros sp., 200
Africa, East and Central, 129–204
  North-West, 116–27
  South, 162, 215–81
  Southern, 206–81
Afrochoerus sp., 117
age variation, skeletal, 6–8, *7*
Alcelaphus bubalis, 124
  helmei, 279
Alcephalini, 200
Alces latifrons, 56
Algeria, 116–22
alveolar processes, 40, 48, 176, 210, 247, 295–6, 316
aminoacid racemization, dating by, 137
amphibia, 135
Amud I–IV, 103, 105, 111
  man, 103
  remains, 100, **103–8**, 126
Anatomy, **3–14**
antelopes, 117, 261
Anthropopithecus erectus, 285
'Ape-man', 216, 223, 235, 243, 260
Arago XXI (skull), 47
  man, 47
  remains, **47–9**
Arctomys marmoti, 29
arm bones, 194
artefacts, 224, 235–6, 278. *See also* Tools and names of tools
artiodactyls, 208, 217, 224, 261, 274
ashes, 73, 315
Asinus africanus, 124
Atlanthropus mauretanicus, 122
auditory meatus, external, 92, 295
Aurignacian culture, 29, 33, 39, 41, 44, 109
Australoid affinities of Wadjak man, 311–12
australopithecines, affinities of, **269–72**
  gracile, 269–71
  osteodontokeratic culture of, 261
  reclassification of, 300
  robust, 269–71
Australopithecus, 180
  africanus, 162, 216, 269–71
    africanus, 216
    (*sensu strictu*), 199
    transvaalensis, 223, 260

  boisei, 132, 136–7, 150, 165, 190
  capensis, 243
  cf. africanus, 190
  cf. boisei, 150, 163, 167, 190
  (*Paranthropus*) *crassidens*, 24
  prometheus, 260, 269
  robustus, 165, 199, 235, 246, 252–3, 271
    crassidens, 243
    robustus, 235
  sp., 167, 190
  transvaalensis, 223
  (*Zinjanthropus*) *boisei*, 132, 149–50, 163
avian remains, 97, 104, 200, 208
awls, bone, 73
axes, 117. *See also* Hand-axes
Axis lydekkeri, 286, 292

banteng cattle, 302
barytes, 123–4
Basal Member, 199–200
basalt flow, 104, 163, 191–2, 199
bats, 314
beech leaves, 78
biasterionic breadths, 22–3, 45, 80
bicondylar breadths, 57, 87, 93–4, 106–7, 121, 288, 316
bifaces, 199, 244
  handaxe-like, 164
bipedal gait, evidence of, 142–3, 145, 211, 265, 272, 288
Bison priscus, 29, 39, 56, 61
blade tools, 27, 84
boar, wild, 90
bolas stones, 208, 274
bone, worked, 41, 73, 208, 274
bones: 3–8
  accumulation of, 261
  See also under names
boring tools, 134
Bos, 97
  ibericus, 124
  primigenius, 19, 47, 61, 67, 73, 124
  sp., 34, 84, 90, 104
bovids, 19, 135, 160, 164, 170, 191, 200
brachycephalic skulls, 67
brain, endocast, 218, 245
  impress of convolutions, 287, 294
breccia, 43, 47, 72, 90, 96–7, 159, 216, 224, 235, 244, 247, 251, 261, 263, 292, 310, 314
Broken Hill man, 207–8, 212
buffalo, 191
burins, 97, 104

calcanei, 29–31, 69

337

Index

Index

Index

gait, 30, 35
Gallilee remains, 94
Gauss Normal Epoch, Kaena or Mammoth
    event in, 200
Gazella, 97, 261
  atlantica, 124
  cuveiri, 124
  dorcas, 124
  sp., 84, 90
  sp., subgutterosa, 104
  wellsi, 224
genioglossal fossa, 192, 248, 253
geomagnetic polarity, dating by, 137, 200
geomagnetic reversal polarity time scale,
    170
geomorphological dating methods, 218,
    225, 245, 262
gerbils, 135
Germany, 50–65
Gigantopithecus, 297
giraffes, 261
giraffids, 135, 160, 164, 170, 200
glabellae, 145, 195, 197, 218, 226, 247, 297
glenoid fossae, 236, 247, 307
Glossary, 327–33
Gorgopithecus major, 235
Gottweiger Interstadial of Würm Glacia-
    tion, 67, 90
gravels, 19, 56, 60, 97, 104, 190, 199, 305,
    314–15
  Steinheim (layers named), 61
Great Britain, 18–26
Greece, 72–6
Grotte des Infants skeletons, 41
Günz Glaciation, 286
Günz-Mindel Interglacial, 67, 217, 225, 261
Guomde Formation, 167–8, 174
Gypsorhynchus dorti, 217
  minor, 217

Hadar Formation, 199
Hadar remains, 162, **199–204**, 271
Hagenstad variation, 279
hand bones (*see also under names*), 29, 35–6,
    69, 92, 106, 138, 141, 148, 194, 201
hand-axes, 19, 84, 117, 134, 164, 224, 274
hearths, 97
Heidelberg man, 55, 63
  mandible, 49, **55–9**, 74–5, 275, 316
herbivores, 135
Herpestes ichneumon, 208
  mesotes, 235
Hexaprotodon ngandonensis, 306–7
hip bones (innominate bones), 69–70,
    146–7, 180, 201, 225, 228–9, 232, 245,
    249–50
hippopotamids, 135, 170, 191, 200
Hippopotamus amphibius, 84, 90, 117, 274
  gorgops, 135
Hippotragus broomi, 224

Holocene Period, 191, 279
Hominidae, genus and species indet., 243
hominids, fossil, **18–324**
  presentation scheme explained, 16
Homo (genus), 180, 271
  (*Africanthropus*) *helmei*, 278
  africanus, 132, 223
  antiquus, 51
  cf. sapiens, 19, 60
  erectus, 23, 49, 58, 63, 74, 80–1, 110, 122,
     132, 151, 162, 180, 213, 243, 285,
     291, 300
    erectus, 285, 289, 291
    javensis, 285, 316, 319
    pekinensis, 289, 313, 319, 322
    seu sapiens, 77
  ergaster, 167
  (*Euranthropus*) *heidelbergensis*, 53
  europaeus, 51
  florisbadensis, 278
    (*helmei*), 280
  habilis, 132, 139–40, 150
  heidelbergensis, 55
  incipiens, 51
  (*Javanthropus*) *soloensis*, 305
  leakeyi, 132
  marstoni, 19
  mousteriensis, 51
  neandertalensis var. krapinensis, 66
  neanderthalensis, 27, 29, 33, 51, 53
    soloensis, 305, 308
  palaeohungaricus, 77
  '*praesapiens*', 43
  primigenius, 51
    asiaticus, 305, 308
    var. Krapinensis, 66
  (*Protanthropus*) *steinheimensis*, 60
  rhodesiensis, 207, 212
  saldanensis, 273, 276
  sapiens, 23, 41, 45, 63, 70, 94, 100, 111,
    190, 210, 212–13, 278, 280, 308
    fossilis, 280
    neanderthalensis, 27, 31, 33, 36, 51, 100,
     308
    protosapiens, 19
    rhodesiensis, 207, 273, 276, 308
    sapiens, 96, 100
    soloensis, 305, 308
    steinheimensis, 19, 60
    wadjakensis, 310, 312
  soloensis, 299, 305, 308
  sp., 159, 162, 167, 199, 313
  sp. indet., 243
  steinheimensis, 60
  swanscombensis, 19
  transvaalensis, 216, 223, 235, 243, 260
  wadjakensis, 299, 310, 312
Homoioceras baini, 208, 274
Homotherium latidens, 117
Hopefield man, 273
Humbu Formation, 163–4

340

Index

humeri, 7, 29–30, 36, 40, 52, 69, 87, 91, 93–4, 173, 176, 201, 202, 209, 211, 225, 228, 232, 236, 238, 240–1, 245, 262, 265, 317–18, 322
Hungary, 77–81
Hyaena brunnea, 245
  dispar, 245
  makapani, 261
hyaenids, 170
hybridization, 94, 100
hypotrochanteric fossa, 173, 175–6

Ighoud 1–3, 123
  man, 123
ilia, 7, 147, 209, 228–9, 262, 264–5
incisors, 7–10, 34, 57, 68, 86–7, 99, 106–7, 120, 138–9, 164–5, 172, 174, 178, 180, 192, 196, 201, 202, 211, 219, 227–8, 237–8, 248–9, 296, 317, 321
infratemporal fossa, 246
innominate bones. *See* Hip bones
insectivores, 217, 224, 314
ischia, 7, 147, 180, 228–9, 250, 262, 264
Israel, 83–103

Jaramillo Event, 164
Java, 284–312
  man, 49, 58, 203, 285, 289
Jebel Ighoud remains, 75, 111, **123–7**
  Kafzeh remains, 94, **96–102**, 107, 111
  Qafsa man, 96

Kabuh beds, 286, 292–3, 305
Kabwe man, 207
Kada Hadar Member, 199
Karari industry, 169
KBS tuff, 167, 169–71
Kenya, 166–89
Kibish Formation, 190–1, 194
knives, backed, 97
'knuckle-walkers', 194
Kobus sp., 200
Koobi Fora Formation, 167–71, 175, 177
  Ileret, 167, 171–4
  Lower Member, 167, 175–6, 178, 180
  Upper Member, 167, 172–4, 176, 178, 180
  remains (East Rudolf hominids), 151, 162, **166–89**, 201, 203, 271
  list of, 182–6
Krapina man, 23, 66–70, 94
  remains, **66–71**
Kromdraai remains, **235–42**, 245, 269–70
Kubi Algi Formation, 167–8, 171
La Chapelle-aux-Saints, 1, 27
  skeleton, **27–32**, 33–6, 53, 110, 127

lacustrine, 104, 133, 199
Laetolil Beds, 159–60
  remains, **159–62**

La Ferassie 1 to 6, 33–4, 126
  skeletons, **33–8**, 110
lagomorphs, 135, 160, 314
lambda and lambdoid region, 78–9, 125
Lantian man, 313
  1 and 2, 313
  remains, **313–24**
Leecyaena forfex, 244
Leo spelaeus, 73
Leptailurus hintoni, 208
Leptobos, 293
Lepus sp., 19
Levallois technique of tool-making, 97, 134
Levalloiso-Mousterian culture, 74, 90, 97, 100, 124
Libytherium, 164
limestones, 27, 47, 51, 83, 89–90, 97, 103–4, 123, 208, 216–17, 224, 235–6, 244, 260–1, 273, 310, 314
linea aspera, 173–4, 176
lion, 56
loess, 56, 60, 78
long bones, 91, 107
  phases of maturation of, 6–9, 7
  See also under names
Loxodonta africana, 170, 208
  atlantica, 117, 235, 274
lunate bone, 317–18, 322
Lycyaena nitidula, 244
  silbergi, 224, 244

Macacus robustus, 315
Magdalenian culture, 41, 44
Makapansgat remains, 217–18, 225, **260–8**, 269–70
Mammuthus primigenius, 34, 39, 61
  trogontherii, 61
  trogontherii-primigenius, 61
'Man-ape', 214, 223, 235, 243, 260
mandibles, 8, 29, 34, 48, 56–7, 67, 84–6, 98–9, 106, 117–20, 122, 144, 161, 163–4, 172, 174, 178, 192, 201, 218–20, 225–6. 245–8, 253–4, 311, 316, 318
  fragmented, 91–2, 138–9, 193–4, 196, 202, 225, 236–7, 246, 253–4, 262–4, 275, 297, 301
  juvenile, 124–5, 138–40, 160–1, 174, 176, 236, 238, 245, 263
mandibular angles, 30, 49. 56–7, 87, 93, 121
  dimensions, 49, 57, 69, 87, 93–4, 121, 195–6, 202, 231, 239, 255, 257, 265, 275. 298, 320–1
  notches, 34, 48, 56
  tori, 120, 122, 160–1, 176, 248, 253, 316
Marmotta marmotta, 67
Masek Beds, 133–4
mastoid processes, 29, 34, 62, 67, 70, 86, 92, 98, 106, 138, 146, 176, 195, 209, 226, 236, 294, 307, 316
Mauer man, 55, 58
  mandible, **55–9**

341

Index

Index

Index

Index

Solo man, 23, 63, 111, 212–13, 275–6, 299, **305–8**
Solutrean Period, 41
South Africa, 160, 215–81
South African stone age chronology, 274
Spectrum hypothesis, 110–11
Spee, curve of, 174, 179
sphenoid bones, 48, 92, 145, 218, 236
spheroids and subspheroids, 224, 244
stalactite, 73
stalagmite, 43, 47, 72–3, 96
stance, 35
stature, 36
Stegodon, 286, 292, 307
Steinheim calvaria, **60–5**
  man, 23, 60, 110
  skull, 45, 60–5
Sterkfontein Extension Site, 223–5
  remains, 201, 203, 217–20, 223–4, 252, 269–70
  skull, 226–7, 263–4
  Type Site, 223–5
sternum, 92
Stillbay culture, 208, 274
Stone Age culture, Early, 274
    Late, 199–200, 274
    Middle, 199, 274, 279
  balls, 306
  implements and tools, 47, 66, 73, 90, 97, 117, 124, 133–5, 164, 169, 208, 224, 244, 274, 279, 306. *See also* Tools *and under names*
'stooping man', 212
Stylochoerus nicoli, 135
Stylohipparion albertense, 135
  steytleri, 235
Suids (pigs), 160, 164, 170, 191, 200, 261. *See also* Sus
Suncus, cf. etruscus, 235, 244
  etruscus, 224
Superciliary ridges. *See* Supraorbital ridges
supramastoid crests, 138, 146, 171, 173, 178, 209, 294, 297
Supraorbital ridges (tori), 29, 34, 40, 48, 52, 62–3, 67, 70, 74, 85, 92, 98, 106, 124–5, 145–6, 169, 175, 180, 209, 226, 236, 247, 253, 275, 280, 287, 293, 297, 307, 311, 319
  visor, 316
Supratoral sulcus, 48, 124, 297
Sus gadarensis, 84, 90
  lydekkeri, 315
  macrognathus, 306
  scrofa, 47
  sp., 34, 104
  terhaari, 306–7
sutures, 22, 52, 98, 121, 125, 138, 287
Swanscombe man, 19, 41, 64, 80, 110–11
  skull fragments, 18, **19–26**, 45, 63
  tools, 19

Swartzkrans remains, 201, 203, 219, 225, **243–59**, 264–5, 269–70
Sylvian crest, 121
Symphyses and symphysial regions, 48, 56, 86, 99, 106, 120, 160, 164, 172, 176, 178, 192, 202, 218, 227, 248, 253, 264, 316. *See also* Mandibles; Mandibular
syphilis, 53

Tabūn I (C1) and II (C2), 83–8, 90, 92–4, 107, 110–11, 127
  remains, **82–8**, 90, 92
tali, 29, 31, 69–70, 174, 201, 236, 239, 311
Tanzania, 132–65
'*Tapinochoerus*' *meadowsi*, 224, 245
Tapirus indicus, 315
tarsal bones, 142–3
Taung skull, **216–21**, 262, 269–70
taurodontism, 57, 63, 69–70, 317
Tautaval man, 47
Tayacian Period, 43, 84
teeth, **8–12**, *13*, 34, 48, 56–8, 62–3, 67–70, 80–1, 86–7, 92–4, 98–9, 105, 120, 125–6, 137–40, 144, 148–9, 160–1, 172, 174, 176, 178, 194, 196–7, 211, 218–21, 225–8, 232, 237–8, 240, 245–9, 255–7, 264–5, 296–8, 301, 311, 314, 316–19, 321–2. *See also* Tooth *and under names*
  anteroposterior disproportion in, 164, 172, 192
  deciduous, 8–9, *13*, 78, 125–6, 160–1, 174, 176, 178, 219–21, 228, 232, 236, 238, 240, 245, 249, 255–6, 317, 321
  modified for omnivorous diet, 272
  unerupted, 125, 139, 176, 178, 237–8
Telanthropus I (restored), 246, 253–4, 257
Telanthropus capensis, 243
temporal bones, 138, 144–5, 173, 195, 201, 236, 246, 293–5, 319
  fossa, 218
  lines, 287, 294
temporomandibular joints, 92, 237
Ternifine man, 49, 58, 117
  remains, **117–22**
Testudo graeca, 44
  sp., 47
Thabaseck Travertine, 217
Thallomys debruyni, 217
Thenailurus barlowi, 224, 261
Theropithecus, 136, 160
thumb bones, 36, 92, 141, 245, 249, 256
tibiae, 7, 29, 31, 35–6, 40, 69, 86–7, 93–4, 107, 145, 174, 176, 194, 201, 209, 307, 311, 318
toe bones, 240
  great, 143
tools (*see also* Stone implements and tools *and under names*)
  Acheulian, 19, 47, 66, 84, 134, 164, 274
  Aurignacian, 39, 41, 44
  Clactonian, 44

Index

tools—*cont.*
 Levalloiso-Mousterian, 84, 90, 97, 100,
  104, 124
 Magdalenian, 41, 44
 Mousterian, 27, 29, 34, 44, 66, 90, 100
 Palæolithic, Upper, 104
 'pre-Aurignacian', 66
 Solutrean, 41
 Tayacian, 44, 47, 84
 Yabrudian, 84
tooth attrition, 7, 34
 crown morphology, 9–10, 86
 eruption, ages of, 12, *13*
 morphology, **8–13**
 See also Teeth
Tragelaphus, 160
 cf. nakuae, 200
trapezium, 141
travertine deposits, 77, 224
 Thabaseck, 217
Trinil 1–9, 285
 calotte, 285–7, 294
 deposit and fauna, 286, 292–3
 remains, **285–90**, 293, 299
trochlear notch, 147
Trogontherium schmerling, 78
tympanic bone, 237, 247
 plate, 295

ulnae, 7, 29–30, 52, 69, 86–7, 93–4, 147,
  174, 176, 193, 196, 201, 236, 238
underbite, 34
ungulates, 286, 292
upper limb girdle, 194
upright posture, evidence for, 252, 265,
  272, 288, 294
uranium fission track, dating by, 104–5
uranium/ionium growth, dating by, 104
Ursus arctos, 61

sp., 44, 56
spelaeus, 19, 47, 73
Usno Formation, 190–2, 200

valgus deviation, 149
vascular markings (on bones), 22, 40, 56
vaults, cranial, 34, 40, 44–5, 48, 52, 62–3,
  67, 74, 85, 92, 98, 105–6, 121, 125, 138,
  141, 144–6, 173, 175, 178, 180, 194–5,
  209, 226, 275, 280, 287, 294, 296–7,
  311, 319
venous sinuses, 80
vertebrae and vertebral column, 29–30,
  69, 86, 194, 201, 225–6, 245, 251–2
Vertésszöllös man, 77
 remains, 23, 74–5, **77–81**
Vieillard, le, 39
Villafranchian Period, 215. *See also*
  Pleistocene Period, Lower
Vitamin D deficiency, 53
Volcanic materials (including tuffs), 133,
  159–60, 163, 167, 169, 190, 199, 285–6,
  292–3

Wadjak man, 299, 308, 310
 remains, **310–12**
wolf, 56
Wolff's law, 3
Wormian bones, 79
Würm Glaciations, I, II (main) and III, 29,
  34, 40, 52, 67, 73, 84, 90, 97, 104, 111

Yabrudian Culture, 84
Yugoslavia, 66–71

Zambia, 206–14
Zinjanthropus boisei, 132, 150
zygomatic bones, 48, 92, 171, 201, 210,
  225, 236, 296–7